Bibliografische Information der Deutschen Nationalbibliothek

Die Deutsche Nationalbibliothek verzeichnet diese Publikation in der
Deutschen Nationalbibliografie; detaillierte bibliografische Daten sind
im Internet über http://dnb.d-nb.de abrufbar.

ISBN 978-3-8325-2941-3

Logos Verlag Berlin GmbH
Comeniushof, Gubener Str. 47,
10243 Berlin
Tel.: +49 (0)30 42 85 10 90
Fax: +49 (0)30 42 85 10 92
INTERNET: http://www.logos-verlag.de

Magnetic resonance methods to enable motion robust thermography in abdominal organs during transcutan ablations

Dissertation
der Mathematisch-Naturwissenschaftlichen Fakultät
der Eberhard Karls Universität Tübingen
zur Erlangung des Grades eines
Doktors der Naturwissenschaften
(Dr. rer. nat.)

vorgelegt von
Antje Kickhefel
aus Wolgast

Tübingen
2011

Tag der mündlichen Prüfung: 19.07.2011
Dekan: Prof. Dr. Wolfgang Rosenstiel
1. Berichterstatter: Prof. Dr. Dr. F. Schick
2. Berichterstatter: Prof. Dr. H. Quick

Abstract

Otto Stern and Walther Gerlach showed that atoms in an outer magnetic field only have discrete states for the magnetic moment. They provided the basis for the Magnetic Resonance (MR) experiments to analyze the structure of molecules. The gain in influence of MR in medical science for diagnostic and treatment is related to the countless opportunities to use the MR, which also allows for completely non-invasive estimation of temperatures within the human body. The standard method to monitor temperatures within a human requires the acquisition of two MR images at two different time-points, which leads to unavoidable errors. One image of the native human body to acquire the reference (background) and one image from the heated human body are necessitated. In this thesis a physically and mathematically motivated method is presented to calculate temperature information using a single MR image. The development of this reference-less temperature monitoring method requires a detailed analysis of the standard temperature monitoring. This analysis has to comprise non-moving organs as well as moving organs within the human body. Patient data from the liver was acquired simultaneously with Laser Induced Thermal Therapy. Temperature data in and around the thermal zone was analyzed using an exponential fit based on the solution of the Bioheat equation proposed by Pennes.

In homogeneous tissue with linear changing susceptibility the phase of the MR signal can be described as harmonic function. It is well known, that the solution of the Laplace equation is a harmonic function. Further, artifacts like heating can be described as a non-harmonic function of the MR phase. In conclusion, the harmonic background phase can be calculated by the placement of a border around the heated zone. This border has to be free of heating, within homogeneous tissue and free of artifacts to ensure the harmonic behavior. The boundary value problem can be solved using the two dimensional Dirichlet principle. Using the harmonic behavior of the MR phase and the Laplace equation, a reference-less Proton Resonance Frequency (PRF) method can be presented in this thesis, which enables non-invasive temperature calculation on one phase image acquired during heating using MR within humans. In comparison to the well known standard reference temperature monitoring method, no reference background image has to be acquired and interscan motion artifacts and artifacts induced by swelling / shrinking of tissue are eliminated. The reference-less PRF method enables fast scanning in free breathing and allows for more flexibility than the standard reference PRF method does. Slice orientation and position, number and position of applicators, and size and position of the region of interest can be varied during intervention. For temperature calculation using this reference-less PRF method, a thin, open, and circular border has to be placed around the heating target area.

Reference-less online temperature monitoring was tested on two patients without breathing triggering or gating simultaneously with a thermo therapy of the liver. The results show that this method can be used to predict necrosis and as a safety tool during local thermal therapies. Additional, an explicit comparison was performed off-line between the standard-reference method and the novel reference-less method. For this purpose, the two methods were systematically applied to identical raw-data sets. As expected, in a non-heated ROI within the liver the reference-less PRF method shows a better stability and robustness than the standard reference PRF method.

The reference-less method presented in this thesis is a promising post-processing tool for temperature monitoring of moving and deforming organs. In comparison to the standard reference-subtraction method, the reference-less approach is more flexible for clinical use, eliminates the need for respiratory triggers, hence enables higher temporal resolution, and eliminates inter-scan motion artifacts as well as under- or overestimation of temperature induced by swelling/shrinking of heated tissue.

v

Zusammenfassung

Otto Stern und Walther Gerlach zeigten, dass die magnetischen Momente von Atomen im äuSSeren Magnetfeld diskrete Zustände einnehmen und begründeten damit die Basis der Magnetischen Resonanz (MR). Die unzähligen Möglichkeiten zur Nutzung der MR (z.b. in der Spektroskopie) führten zu einem stetig wachsenden Einfluss der MR, auch in der Medizin. Das Magnetresonanz Verfahren kann als diagnostisches und auch als therapeutisches Verfahren eingesetzt werden, sogar eine nicht invasive Temperaturmessung im menschlichen Körper ist möglich.

Das Standardverfahren zur Temperaturmesung mit dem MR Tomographen benötigt zwei MR Bilder zu zwei verschiedenen Zeitpunkten. Es wird ein MR Bild vom nativen menschlichen Körper als Referenz (Hintergrund-Bild) und eines von demselben Körper im geheizten Zustand benötigt. Der zeitliche Versatz der zwei MR-Bilder führt zu unvermeidlichen Berechnungsfehlern. In dieser Arbeit wird ein physikalisch und mathematisch motiviertes Verfahren zur referenzfreien Temperaturmessung vorgestellt, das nur ein einziges MR-Bild benötigt, um die Temperatur in einem aufgeheizten Organ zu bestimmen.

Die Erforschung dieser neuartigen referenzfreien Methode zur Temperaturmessung setzt eine detaillierte Analyse des Standardverfahrens zur Referenz-Temperaturmessung voraus. Diese Analyse muss sowohl bewegte als auch unbewegte Organe des menschlichen Körpers berücksichtigen. Im Rahmen der vorliegenden Arbeit wird daher das menschliche Gehirn als unbewegtes Organ und die Leber als bewegtes und sich gleichzeitig verformendes Organ analysiert.

Um den Einfluss der Bewegung auf die Temperatur-Daten zu minimieren, wird in dieser Arbeit eine Nachverarbeitung auf der Basis einer Wärmeleitungsgleichung, der sogenannten „Bioheat" Gleichung von Pennes, erforscht. Sie ermöglicht, nicht nur einen einzelnen Zeitpunkt, sondern jeweils mehrere Zeitpunkte bei der Darstellung der Temperatur zu berücksichtigen.

In homogenem Gewebe mit einer linear variierenden Suszeptibilität kann die Phase des MR-Signals als harmonische Funktion beschrieben werden. Artefakte, wie z.B. beim Erhitzen von Gewebe, können als nicht-harmonischer Anteil einer Funktion verstanden werden. Die Lösung der Laplace Gleichung ist eine harmonische Funktion. Die harmonische Hintergrundphase kann daher berechnet werden, indem man eine geschlossene Fläche um den geheizten Bereich legt. Diese geschlossene Fläche darf weder geheizt werden, noch andere Artefakte enthalten, die ein nicht-lineares Verhalten der Suszeptibilität verursachen. Weiterhin muss diese Fläche in homogenem Gewebe platziert sein, damit der harmonische Charakter der darauf liegenden Phase gewährleistet werden kann. Dieses Randwert-Problem kann dann mit dem Dirichelet Verfahren gelöst werden. Nutzt man das harmonische Verhalten der MR-Phase in homogenem Gewebe und gleichzeitig die Laplace-Gleichung, entsteht daraus die Möglichkeit die Temperatur nicht invasiv mit dem MR Tomographen referenzfrei aus einem einzelnen MR-Bild zu bestimmen. Im Gegensatz zum Standard-Temperatur-Messverfahren muss kein Referenz-MR-Bild vor der Thermotherapie aufgenommen werden, wodurch Artefakte durch die Bewegung zwischen den Aufnahmen und durch das Anschwellen und Schrumpfen von Gewebe eliminiert werden. Die in dieser Arbeit vorgestellte referenzfreie Temperaturmessung mittels MR ermöglicht eine Therapieüberwachung ohne Atemtriggerung. Es kann der gesamte Atemzyklus zur Temperaturdarstellung genutzt werden, wodurch die Frequenz der Aufnahmen um ein Vielfaches gesteigert werden kann. Weiterhin erhöht sich die Flexibiltät während der Intervention. Es können Schichtorientierung, -angulierung und -positionierung sowie Größe und Position des zur Temperaturmessung verwendeten Bereichs während der Intervention verändert werden. Darüber hinaus können Anzahl und Position der Applikatoren individuell während der Behandlung angepasst werden.

Die Online-Darstellung der referenzfreien Temperaturmessung wurde am MR ohne Atemtriggerung während einer Intervention an der Leber von zwei Patienten getestet. Das Verfahren ermöglicht sowohl die Abschätzung der entstehenden Nekrose als auch eine Echtzeit-Sicherheits-Überwachung der Thermotherapie. Weiterhin wurde ein direkter Vergleich zwischen der Standardmethode und der neuen referenzfreien Methode durchgeführt. Die Berechnung der Temperaturen erfolgte offline mit denselben Rohdaten für beide Verfahren. Dabei zeigt die referenzfreie Temperaturmessung im ungeheizten Bereich eine bessere Stabilität und Genauigkeit als die Standard-Methode.

Die in dieser Arbeit präsentierte referenzfreie Temperaturmessung ist eine viel versprechende Nachverarbeitungsmethode zur Darstellung der Temperatur an sich bewegenden und verformenden Organen. Lokale thermale Therapien an bewegten Organen, wie der Leber, können mit der vorgestellten nicht invasiven referenzfreien Methode zur Temperaturmessung überwacht werden. Im Vergleich zur Standard-Referenz-Subtraktions-Methode ermöglicht die referenzfreie Methode mehr Flexibilität in der klinischen Anwendung, den Verzicht auf Atemtriggerung, eine höhe zeitliche Auflösung der MR-Messungen. Die unvermeidlichen Bewegungs-Artefakte zwischen den Aufnahmen, sowie die Über- oder Unterschätzung der Temperatur durch das Anschwellen oder Schrumpfen des geheizten Gewebes, wie sie bei der Verwendung der Standard Referenz PRF-Methode auftreten, können mit der vorgestellten referenzfreien Methode verhindert werden.

Contents

Introduction

An interesting experiment has been made by Otto Stern and Walther Gerlach in 1921. They took a beam of electrically neutral silver atoms and let it pass through a non-uniform magnetic field. This magnetic field deflected the silver atoms like they would deflect little dipole magnets if you threw them through the magnetic field. They showed that atoms have only discrete settings for the magnetic moment [45, 46]. In 1938, experiments of Rabi determined the magnetic moment. Both studies, the one of Rabi and the one of Stern and Gerlach, were the basics for the examinations of magnetic resonance performed by Bloch and Purcell. In 1946 both published independently the first successful Magnetic Resonance (MR) experiments in fluid and solid phase [108]. Nearly 30 years later in 1973, Lauterbur first enabled spatially encoded NMR measurements [72]. Peter Mansfield developed methods that have improved the resolution and speed of MRI to such an extent that images can now be captured in a matter of seconds, not hours [83, 84].

In parallel to the described experiments the theoretical basics were set up. In 1928, Dirac formed the Dirac equation on the basis of the exclusion principle of Pauli [35, 36]. The Dirac equation relies on wave equations of the special theory of relativity [37]. This Dirac equation is a theoretical description of the quantum mechanical properties and behavior of fermions (Spin 1/2 particle) and the theoretical explanation of the Stern-Gerlach experiment. Together with Enrico Fermi, Dirac formulated the Fermi-Dirac statistics to describe the behavior of fermions. This Fermi-Dirac statistic is the theoretical basic of all MR experiments.

For a system of identical fermions, the Fermi-Dirac distribution correlates the temperature and the averaged number of fermions in a single-particle state [63]. This correlation is the fundamental basis of temperature monitoring using NMR.

The minimally invasive thermal therapy is a rapidly-developing local treatment of solid tumors [14,18,39,55,56,60,79,80,95,97,107,119,137,138]. Local thermal therapies have several advantages compared to surgery, chemotherapy, radiotherapy, and immunotherapy. Thermal ablations allow for benefits such as the anticipated reduction in morbidity and mortality, minimal side effects, and local application. Local thermal therapies can be performed under real-time imaging guidance [109, 112]. Outpatient procedure can be provided and a wide spectrum of patients including those who are not surgical candidates can be treated. Another essential advantage of thermal ablation is that the treated area has no "thermal memory", which enables the repeated thermal ablation in the same spot. Also, the results of thermal ablation on solid tumors are immediate, whereas chemotherapy and radiation take more time to shrink tumors. The goal of local thermal therapies is to heat diseased tissue to induce necrosis. At the same time, the surrounding healthy tissue has to be below toxic levels. The spatial-temporal delivery of heat energy requires control which can be verified using the temperature dependent parameters of NMR. The preferred NMR parameter is the Proton-Resonance-Frequency shift (PRF).

Chapter 1 gives a brief overview of the fundamentals. First, the basics of Magnetic Resonance Imaging (MRI) and the design of MRI-sequences (Section 1.1.4) are explained, and the GRadient Echo (GRE) sequence (Section 1.1.5) and the Echo Planar Imaging (EPI) sequence (Section 1.1.6) are discussed. Second, temperature monitoring for clinical practice is explained and requirements for online temperature visualization (Section 1.2) are defined. Third, several temperature dependent MRI parameters are discussed. It is described how these parameters can be used for temperature monitoring. Additionally, advantages and disadvantages of each parameter are discussed (Section 1.3). Section 1.4 describes heat transfer in biological tissue and dosimetry estimation. Section 1.1 gives a brief introduction to MRI and GRE based pulse sequences.

The aim of this thesis is the improvement of magnetic resonance temperature monitoring for clinical practice, particularly on moving and deforming organs. Chapter 2 and 3 are used to analyze in detail the standard reference PRF method to identify prospects, restrictions and the scope for development. The analysis has to be divided into temperature monitoring at non-moving organs (Chapter 2) and at moving organs (Chapter 3).

In Chapter 3 online temperature monitoring was examined using the standard reference PRF method of patients' livers during an intervention. Temperature monitoring was performed using a breathing belt to administer two MRI acquisitions on the liver, twice in the same slice position / orientation and with the same deformation of the liver. In spite of breath triggering, it could not be guaranteed that each voxel of the liver during the intervention would be in the same position as the corresponding voxel within the reference image.

To improve certainty of temperature monitoring, an exponential fit based on the Bioheat equation proposed by Pennes is developed in section 3.2.3.1. Using the fit, information of each voxel includes information from all the previously acquired data points. Increasing the information content of each voxel also increases the temperature certainty and decreases the dispersion of the temperature values.

Nevertheless, the standard implementation of the standard reference PRF method is based on the subtraction of a temporal reference phase map and is therefore intrinsically sensitive to tissue motion (including deformation) and to external perturbation of the magnetic field. Interscan movement and deformation of organs like the liver are the main cause of over or under estimation of temperature using the standard reference PRF method. A method to overcome this problem is presented in Chapter 4. The reference-less PRF method has been previously described by V. Rieke, based on a 2D polynomial fit [113]. While their implementation was demonstrated to be a major progress in term of robustness against the perturbations mentioned above, the underlying mathematical formalism requires a thick unheated border and may be subject to numerical instabilities. A novel method of reference-less PRF is presented in this thesis, using a physically consistent formalism, which exploits mathematical properties of the magnetic field in a homogeneous or near-homogeneous medium.

The measured MRI phase map less the temperature-induced shift is called the background (or reference) phase. This background phase should be calculated and subtracted from the measured phase map in order to isolate the temperature elevation. Taking a 3D domain with no sources of magnetic susceptibility, the unwrapped MRI phase of the 3D data is demonstrated to be a harmonic function, which results in a null scalar field when the Laplacian operator is applied. Susceptibility inhomogeneities, localized heating, mixed signal from different chemical shift species with spatially varying ratios, or other artifacts, are all non-harmonic functions which contribute to the globally measured MRI phase [122]. Calculating the background phase means calculating the harmonic part of the phase map. The non-harmonic part of the MRI phase, in a homogeneous medium, constitutes the temperature elevation. Phase information along a one pixel thick border is theoretically sufficient to calculate the full harmonic background phase inside the domain (i.e. inner Dirichlet problem). This border has to be unheated and free from artifacts and susceptibility contrast. To enable slice-per-slice monitoring of temperature, the calculation of the background phase was reduced from 3D to 2D with a residual term in the phase's Laplacian of order zero. Moreover, openings in the border were compatible with the method by considering a circular geometry of the domain. The open border implementation permits the elimination of areas of conflict from the assigned border, such as heating, artifacts and susceptibility contrast.

The method has been validated experimentally by comparison with the "ground truth" considered to be the standard reference PRF method for static ex vivo tissue, non-heated volunteers and patient data acquired during a thermo therapy.

Chapter 1

Basic Principles

1.1 Basics of Magnetic Resonance Imaging (MRI)

1.1.1 Magnetic resonance

Magnetic moment μ and total angular momentum J of nuclei are proportional. Both vectors $\vec{\mu}$ and $\vec{J}$ are parallel and related by:

$$\vec{\mu} = \gamma \cdot \vec{J} \qquad (1.1)$$

with the proportionality constant γ, called gyromagnetic ratio. For a given nucleus, γ is constant and characteristic. The spin I of protons and neutrons can be handled as quantum operator and J can be defined by (Figure 1.1):

$$\vec{J} = \frac{h \cdot \vec{I}}{2\pi} = \vec{I}\hbar \qquad (1.2)$$

$$|\vec{\mu}| = \gamma\hbar\sqrt{I(I+1)} \qquad (1.3)$$

$$\mu_z = \gamma\hbar m \qquad (1.4)$$

where $h = 2\pi\hbar$ is Planck's constant. Only the z-component of $\vec{I}$ can be measured. The quantum number m of I_z can only be an integer- or half-integer-value in intervals of 1 from $-I$ to I. Protons, electrons and neutrons have $I = 1/2$. Bringing a nucleus into a static magnetic field $\vec{B_0}$, the energy of the nucleus is minimal if the magnetic dipole momentum is parallel to $\vec{B_0}$. If the magnetic momentum and the main magnetic field are not parallel, an angular momentum appears and the spins precess around the direction of the main magnetic field. The interaction energy of the nucleus is described by the so called Zeeman Hamiltonian:

$$H_z = \gamma \cdot B_0 \cdot \hbar \cdot I_z \qquad (1.5)$$

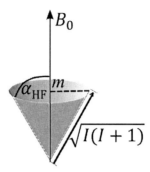

Figure 1.1: *Visualization of the spin $\vec{I}$ in an eigenstate. B_0 is defined by the macroscopic magnetic field, α stands for the flip angle, $\vec{I}$ is the spin of a proton, and m the quantum number of I_z (z-component of $\vec{I}$).*

H_z and I_z are proportional and have the same eigenfunction with the following eigenvalues of H_z:

$$E(m) = \gamma \cdot B_0 \cdot \hbar \cdot m \tag{1.6}$$

These eigenvalues are the allowed values of energy for a free nucleus. The Zeeman energy is the energy of spin interaction with the magnetic field. The energy levels are equally spaced. The energy difference between neighboring levels is:

$$\Delta E = \gamma \cdot B_0 \cdot \hbar \tag{1.7}$$

Transition between adjacent energy levels can be induced by applying an additional alternating magnetic field:

$$\vec{B} = \vec{B_0} + \vec{B_1}(t) = \vec{B_0} + \sin(\omega_0 t)\,\vec{e_x} + \cos(\omega_0 t)\,\vec{e_y} \tag{1.8}$$

The energy of the photons of this frequency is:

$$E = \omega \cdot \hbar \tag{1.9}$$

and the resonance occurs when:

$$\omega = \omega_0 = \gamma \cdot B_0 \tag{1.10}$$

where ω_0 is the resonance frequency of the precision called "Larmor frequency".

The spin population difference in the two spin states is related to the energy difference in the

Boltzmann relationship:

$$\frac{n_1}{n_2} = exp\left(\frac{\Delta E}{\kappa T}\right) \tag{1.11}$$

where κ is the Boltzmann's constant, T the absolute temperature and n_1, n_2 are the populations of the two Zeeman levels.

1.1.2 Radio frequency excitation and relaxation

Applying an RF excitation pulse (Equation 1.8) at the Larmor frequency rotates or tippes the net magnetization relative to the main magnetic field by a so called flip angle (Figure 1.1):

$$\alpha_{RF} = \int_0^{t_{RF}} \gamma B_1(t')dt' \tag{1.12}$$

where $B_1(t')$ defines the amplitude of the radio frequency signal. The derivation of the expectation vector $\langle \vec{\mu} \rangle$ with respect to time can be described as:

$$\frac{d\langle \vec{\mu} \rangle}{dt} = \langle \vec{\mu} \rangle \times \gamma \vec{B} \tag{1.13}$$

In practice, a radio number of protons is excited at the same time by a radio frequency pulse. Therefore, it is necessary to introduce the bulk magnetization $\vec{M} = \sum_i \vec{\mu_i}$, which sums over all magnetic moments within a macroscopic volume of the sample. Assuming that the protons do not interact, equation 1.13 is also valid for the bulk magnetization.

After the spins drop, they relax back to the direction of the main magnetic field (Figure: 1.2 and 1.3).

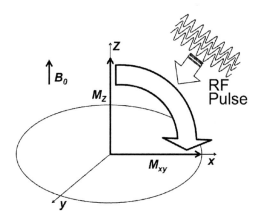

Figure 1.2: Spins are flipped from the direction of the main magnetic field (z-direction) toward the x-y-plane by applying a radio frequency signal.

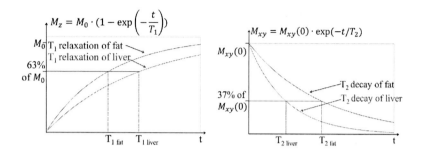

Figure 1.3: *After flipping the spins by a radio frequency signal, two main relaxation processes occur. Adjacent relaxation can be observed in z-direction and in x-y-plane.*
Left: *The T_1-relaxation time characterizes the rate at which the longitudinal M_z component of the magnetization vector recovers after a 90° excitation.*
Right: *The T_2-relaxation time characterizes the rate at which the M_{xy} component of the magnetization vector decays after a 90° excitation.*

To account for the relaxation processes, Bloch extended equation 1.13 by respective relaxation terms, to the so called *Bloch equation*

$$\frac{d\vec{M}(t)}{dt} = \vec{M}(t) \times \gamma \vec{B}(t) + \begin{pmatrix} -M_x(t)/\,T_2 \\ -M_y(t)/\,T_2 \\ (M_0 - M_z(t))/\,T_1 \end{pmatrix}, \tag{1.14}$$

which makes it possible to describe the evolution of the magnetization observed during MRI. In contrast to equation 1.13, the Bloch equation is written for the bulk magnetization and the empirically found relaxation terms.

The *spin-lattice* or *longitudinal relaxation* time is characterized by the relaxation time T_1 and describes an exponential recovery of the longitudinal magnetization M_z (Figure 1.3). The effect is attributed to energy exchange between the protons and their environment, returning the excited system to its thermal equilibrium state M_0:

$$M_z(t) = M_0 \cdot \left(1 - e^{\left(-\frac{t}{T_1}\right)}\right) \tag{1.15}$$

Typical T_1 values at 1.5 T in the human liver are $T_1 \cong 490\,\text{ms}$, in the human brain for gray matter $T_1 \cong 950\,\text{ms}$ and for white matter $T_1 \cong 600\,\text{ms}$ (Table 1.1).
While M_z increases, M_{xy} decreases. The decreasing of M_{xy} is called *spin-spin relaxation* or *transverse relaxation* time and characterized by the relaxation time T_2, which corresponds to a dephasing of the moments inside a macroscopic volume.

Table 1.1: *Typical T_1 and T_2 relaxation times at 1.5 T for several human organs [49].*

Tissue	T_1 [ms]	T_2 [ms]
gray matter	950	100
white matter	600	80
muscle	900	50
fat	250	60
blood	1200	100-200
liver	490	40

Because the bulk magnetization averages over all moments in the volume, its amplitude decays from destructive interference. The effect originates from frequency fluctuations caused by proton interactions:

$$M_{xy}(t) = M_{xy}(0) \cdot e^{\left(-\frac{t}{T_2}\right)} \tag{1.16}$$

Both time constants, T_1 and T_2 relaxation, are properties of the individual sample material. The T_2 relaxation time is always shorter than the T_1 relaxation time.

There is also a reversible bulk field dephasing effect caused by local field inhomogeneities, and its characteristic time is referred to as $T_2^{\star}$ relaxation. These additional dephasing fields come from the main magnetic field inhomogeneity, the differences in magnetic susceptibility among various tissues or materials, chemical shift, and gradients applied for spatial encoding. This dephasing can be eliminated by using a 180° pulse, as in a Spin Echo (SE) sequence. In a SE sequence, only the "true" T_2 relaxation is seen. A GRE sequence does not have a 180° refocusing pulse, so dephasing effects are not eliminated. Therefore, the signal decay in the GRE sequences is a combination of "true" T_2 relaxation and signal dephasing caused by magnetic field inhomogeneities. $T_2^{\star}$ is shorter than T_2 (Section 1.1.5).

1.1.3 2D Fourier reconstruction

The complex values acquired during an MRI measurement are saved within the so called k-space with the coordinates k_x and k_y. MR imaging can be expressed as a two dimensional object given as:

$$S(k_x, k_y) = F\{I(x,y)\} \tag{1.17}$$

where F represents the spatial information encoding scheme. If F is invertible, a data consistent I can be obtained from the inverse transformation:

$$I(x,y) = F^{-1}\{S(k_x, k_y)\} \tag{1.18}$$

The desired image intensity function $I(x,y)$ is a function of the relaxation times (T_1, T_2, $T_2^{\star}$), the spin density, and several other MRI parameters.

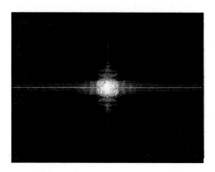 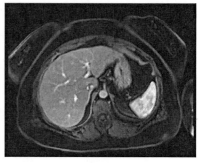

Figure 1.4: Left: *The acquired MR-raw-data $S(k_x, k_y)$ within the k-space have a complex character. The displayed image shows the absolute value of the k-space:*
$\sqrt{real(S(k_x, k_y))^2 + imag(S(k_x, k_y))^2}$
Right: *MR magnitude image $|I(x, y)|$ after Fourier transformation. A human liver is displayed. MR magnitude images are displayed in arbitrary intensity values in this thesis.*

After the slice-selective excitation, the MR signal can be spatially encoded, by applying magnetic field gradients (G_x and G_y) to the remaining two dimensions (Equation 1.26). If it is assumed that the position of the magnetization with respect to the coils does not change with time (patient does not move), the signal can be written as:

$$S(t) = \int_{-\infty}^{+\infty} \int_{-\infty}^{+\infty} I(x, y) \cdot e^{i\omega(x) + i\varphi(y)} dx dy \tag{1.19}$$

$$= \int_{-\infty}^{+\infty} \int_{-\infty}^{+\infty} I(x, y) \cdot e^{i\gamma \int_0^T G_x x dt + i\gamma \int_0^{T_{PE}} G_y y dt} dx dy \tag{1.20}$$

where T is the time for applying the frequency encoding gradient and T_{PE} the time for applying the phase encoding gradient. Making the following substitution:

$$k_x = \gamma \int_0^T G_x(t) dt \tag{1.21}$$

$$k_y = \gamma \int_0^{T_{PE}} G_y(t) dt \tag{1.22}$$

Equation 1.20 becomes:

$$S(k_x, k_y) = \int_{-\infty}^{+\infty} \int_{-\infty}^{+\infty} I(x, y) \cdot e^{ik_x x + ik_y y} dx dy \tag{1.23}$$

This signal, in the so-called k-space, is the 2D Fourier transformation of the spin density distribution, $I(x, y)$, created by applying phase- and frequency-encoding gradients (G_x, G_y). $S(k_x, k_y)$ represents the frequency spectrum of the spin density distribution. An image corresponding to the spatial variation in proton density can be obtained by calculating the inverse two dimensional Fourier transformation of the measured data (Figure 1.4). In most cases the magnitude image (Figure 1.4) is used in the clinical practice. The calculation of relative temperatures is based on the calculation of phase images (Figure 1.5).

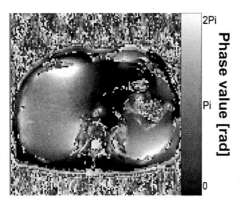

Figure 1.5: *The complex image $I(x,y)$ provides a magnitude image $|I(x,y)|$ as well as a phase image. Writing the complex image as $I(x,y) = x + iy$, where x defines the real and y the imaginary part of $I(x,y)$, the phase image can be calculated as $\phi = arctan(y/x)$. The phase image is used for temperature monitoring.*

1.1.4 Structure of a common MRI pulse sequence

MRI pulse sequences are comprised of Radio-Frequency-pulses (RF-pulses), gradient waveforms, and data acquisition.

RF-pulses can be described using the envelop $B_1(t)$ (Figure 1.6), pulse width (T), bandwidth ($\Delta f \approx 1/t_0$), and the flip angle (α). For a GRE sequence, sinc-pulses are used to apply the excitation. Up to flip angles of 30°, the following relationship is a valid approximation:

$$B_1(t) = \begin{cases} F\left[A \cdot t_0 \frac{sin\left(\frac{\pi \cdot t}{t_0}\right)}{\pi \cdot t}\right] = 1 & : & -N_L \cdot t_0 \leq t \leq N_R \cdot t_0 \\ 0 & : & elsewhere \end{cases} \tag{1.24}$$

F stands for the Fourier transformation, A is the peak RF amplitude occurring at $t = 0$, t_0 is one-half the width of the central lobe, and N_L and N_R are the number of half-periods (zero-crossings) in the sinc-pulse to the left and right of the central peak (Figure 1.6).

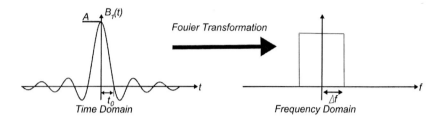

Figure 1.6: *A typical sinc-pulse is shown on the left in the time domain. The envelop of the RF-Pulse $B_1(t)$ is plotted over time. t_0 is one-half the width of the central lobe and A the peak of the radio frequency amplitude at $t = 0$. Both parameter A and t_0 characterize the sinc-pulse. The right side shows the frequency domain and the bandwidth (Δf) of the RF pulse.*

Magnetic field gradients encode spatial information into MR signals. For spatial encoding, three gradients are needed:

$$\vec{G} = \frac{\partial B_z}{\partial x}\hat{x} + \frac{\partial B_z}{\partial y}\hat{y} + \frac{\partial B_z}{\partial z}\hat{z} \tag{1.25}$$

$$= G_x\hat{x} + G_y\hat{y} + G_z\hat{z} \tag{1.26}$$

where $\hat{x}$, $\hat{y}$, and $\hat{z}$ are the unit vectors of the Cartesian coordinate system and G_x, G_y, and G_z are the three orthogonal components of $\vec{G}$ (Figure 1.7).

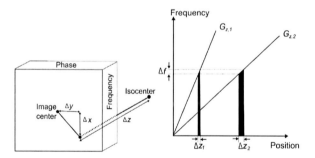

Figure 1.7: Left: *Prescription of a slice plane relative to the isocenter. First the slice encoding gradient G_z is applied to excite a certain slice and afterward the phase- and frequency encoding gradients G_x and G_y are applied to encode in 2D-plane.*
Right: *Plot of Larmor frequency versus position along the direction of the slice-selection gradient. The slope of each of the two lines represents the strength of a slice-selection gradient $(G_{z,1}$ and $G_{z,2})$. For any given bandwidth Δf of the RF pulse, the stronger gradient produces a thinner slice Δz_1.*

The Gradient Echo [43] and the Spin Echo (SE) [52] sequence are the most common sequences, which are used as basic pulse designs to design for example faster sequences like the Echo Planar Imaging (EPI) sequence. The GRE based sequences are a class of pulse sequences primarily used for fast scanning. The echo is formed by a gradient reversal on the frequency-

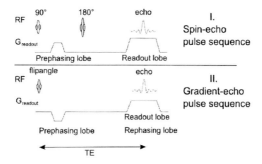

Figure 1.8: *Above:* *Frequency-encoding gradient waveform in a SE pulse sequence. Waveform consists of two lobes; a prephasing gradient lobe and a readout gradient lobe, both have the same polarity. The 180°-pulse is used to form an echo.* *Below:* *Frequency-encoding gradient waveform in a GRE pulse sequence. The two gradient lobes, the prephasing gradient lobe and the readout gradient lobe have opposite polarity. The GRE sequence uses the readout gradient to produce an echo.*

encoded axis. First, spins are dephased using a readout prephase gradient lobe. Second they are rephased using a readout gradient lobe that has opposite polarity. Compared to the GRE sequence, the SE sequence uses a 180°-RF refocusing pulse to form a spin echo (Figure 1.8).

1.1.5 Gradient echo sequence

Figure 1.9 shows a typical pulse sequence diagram for a GRE sequence. The peak of the GRE echo occurs when the area under the two gradient lobes in the frequency encoding direction is equal.

The contrast of the GRE images is weighted by a factor of $e^{\frac{-TE}{T_2^*}}$ instead of $e^{\frac{-TE}{T_2}}$ as in spin echo images. The parameter T_2^* is related to the spin-spin-relaxation time T_2 by:

$$\frac{1}{T_2^*} = \frac{1}{T_2} + \frac{1}{T_2'} \tag{1.27}$$

where T_2' is inversely proportional to the magnetic field inhomogeneity ΔB in each imaging voxel:

$$T_2' \propto \frac{1}{\gamma \cdot \Delta B} \tag{1.28}$$

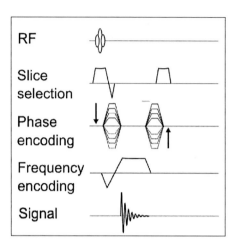

Figure 1.9: *Pulse sequence for spoiled GRE acquisition.*

Each excitation pulse converts longitudinal into transversal magnetization and produces a Free-Induction Decay signal (FID) that can be rephased into a gradient echo. To destroy any remaining transverse magnetization the readout gradient, a spoiled GRE sequences can be used. The spoiler gradient on the slice select axis during the end module, destroys the remaining magnetization, which exists for short repetition times. As a result, only z-magnetization remains during a subsequent excitation. If a perfect spoiling is assumed, $M_{xy}(t_1)$ is zero and $M_z(t_1)$ is maximal just before each RF-pulse. After excitation, the Bloch equation describes the time evolution of magnetization to the external magnetic field and the relaxation times (Equation 1.14):

$$\frac{d\vec{M}}{dt} = \vec{M}(t) \times \gamma \cdot \vec{B}(t) - \frac{M_z(t) - M_0}{T_1} \cdot \vec{e_z} - \frac{\vec{M_{xy}(t)}}{T_2} \tag{1.29}$$

$$\implies \begin{pmatrix} \frac{dM_x(t)}{dt} \\ \frac{dM_y(t)}{dt} \\ \frac{dM_z(t)}{dt} \end{pmatrix} = \begin{pmatrix} \gamma \cdot M_y(t) \cdot B_0 - \frac{M_x(t)}{T_2^*} \\ -\gamma \cdot M_x(t) \cdot B_0 - \frac{M_y(t)}{T_2^*} \\ -\frac{M_z(t) - M_0}{T_1} \end{pmatrix} \tag{1.30}$$

$$\implies M_{xy}(t) = M_{xy}(t_0) \cdot e^{\left(i \cdot \omega t - \frac{t}{T_2^*} \right)} \tag{1.31}$$

$$\implies M_z(t) = M_0 \left[1 - e^{-\frac{t}{T_1}} \right] + M_z(t_0) \cdot e^{-\frac{t}{T_1}} \tag{1.32}$$

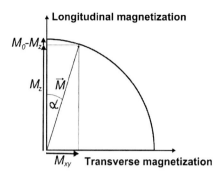

Figure 1.10: *Response of fully relaxed magnetization to a small flip angle RF excitation pulse. The amount of transverse magnetization M_{xy} created by an RF excitation pulse is much greater than the loss of longitudinal magnetization $(M_0 - M_z)$, which allows short TR and rapid acquisition with GRE.*

For steady-state, we can define $M_z(t) = M_z(t_0) = M_z$. The signal S from the spoiled GRE sequence can be described by:

$$S = M_{xy}(t) \quad = \quad M_{xy}(t_0) \cdot e^{-\frac{TE}{T_2^*}} \tag{1.33}$$

$$= \quad M_z(t_1) \cdot sin(\alpha) \cdot e^{-\frac{TE}{T_2^*}} \tag{1.34}$$

$$= \quad \frac{M_0 \cdot sin(\alpha)(1 - e^{-\frac{TR}{T_1}})}{1 - cos(\theta) \cdot e^{-\frac{TR}{T_1}}} e^{-\frac{TE}{T_2^*}} \tag{1.35}$$

Equation 1.35 characterizes the signal reduction due to the T_2^* decay of a tissue (Figure 1.6). To maximize S by optimizing the flip angle, the *Ernst angle* α_E has to be used:

$$\alpha_E = arccos(e^{-\frac{TR}{T_1}}) \tag{1.36}$$

1.1.6 Echo planar imaging sequence

Echo Planar Imaging (EPI) employs a series of bipolar readout gradients to generate a train of gradient echoes [82]. Multiple k-space lines are sampled under the envelop of one free-induction decay. Using this readout mode, EPI allows for very fast imaging. A phase-encoding gradient is used for the spatial encoding of echoes to sample multiple k-space lines. The number of gradient echoes produced following one RF excitation is called Echo Train Length (ETL). The interval between two adjacent echoes is the Echo Spacing (ESP) (Figure 1.11). A Single Shot EPI (ss EPI) sequence acquires all echoes after a single excitation. A Segmented EPI (seg EPI) sequence uses several segments to acquire the whole k-space.

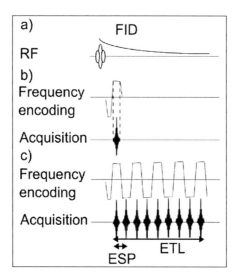

Figure 1.11: a) *Following the RF excitation pulse, the magnitude of the transverse magnetization M_{xy} decays according to T_2^* .* **b)** *First a conventional* **GRE** *acquisition is shown,* **c)** *second an* **EPI** *acquisition. Within the GRE sequence after one excitation pulse one k-space line is acquired. Within the EPI sequence several k-space lines are acquired. The single shot EPI acquires all k-space lines after one excitation.*

The **readout frequency gradient** waveform in EPI starts with a prephasing gradient, followed by a series of readout gradient lobes with alternating polarity. The prephasing gradient area positions the k-space trajectory to $k_x(\min)$. The k-space line spans from $k_x(\min)$ to $k_x(\max)$ (Figure 1.12). The acquisition time for a single echo is given by:

$$T_{\mathrm{acq}} = \frac{n_x}{2 \cdot \Delta\nu} = \frac{2\pi \cdot n_x}{\gamma \cdot L_x \cdot G_x} \tag{1.37}$$

where $2 \cdot \Delta\nu$ is the full receiver bandwidth, n_x is the number of complex k-space data points along the readout direction, L_x is the readout Field Of View (FOV), and G_x is the readout gradient amplitude. Generally, a wide receiver bandwidth results in decreased Signal to Noise Ratio (SNR):

$$\mathrm{SNR} \propto \frac{1}{\sqrt{\Delta\nu}} \tag{1.38}$$

On the other hand, a large receiver bandwidth reduces the effect of T_2^* induced signal decay, which can increase SNR.

The **phase encoding gradient** also starts with a prephasing gradient to determine the initial position of k-space sampling along the phase-encoded direction.

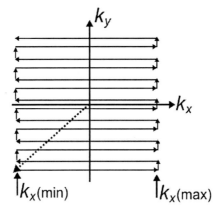

Figure 1.12: *k-space trajectories for the EPI-sequence. The dotted diagonal lines show the trajectory of the prephasing gradients.*

In a gradient-echo EPI, each k-space line along the phase-encoded direction is acquired at a different TE. The amplitude of the corresponding gradient echo decay has the following behavior:

$$S(n) = S_0 \cdot e^{\frac{TE(n)}{T_2^*}} \tag{1.39}$$

where n is the echo index in the echo train and S_0 is the signal at time zero. The effective echo time is defined as the TE that corresponds to the central k-space line.

Compared to conventional GRE imaging, EPI is more prone to a variety of artifacts. A prominent EPI artifact is **ghosting** along the phase encoding direction. System imperfections and physical phenomena can lead to ghosts in EPI (e.g. eddy currents). These ghosts can be reduced or removed using calibration, reconstruction, or post processing techniques. **Chemical shift** artifacts of the first kind are also a well known problem in echo planar images (Figure 1.13). Induced displacement is typically negligible along the EPI readout direction, but becomes severe along the phase encoding direction. For spins with a chemical shift of Δf_{cs} relative to the receiver frequency, the chemical shift produces a spatial shift along the phase encoded direction:

$$\Delta y_{cs} = \frac{\Delta f_{cs}}{\Delta \nu_{phase}} L_y = \frac{t_{esp} \cdot \Delta f_{cs}}{N_{shot}} L_y \tag{1.40}$$

where L_y is the FOV, N_{shot} is the number of required shots to acquire the total k-space for multi shot EPI, $\Delta \nu_{phase}$ is the bandwidth in the phase encoding direction, t_{esp} is the echo spacing. Fat suppression should be applied to reduce chemical shift artifacts.

An **off-resonance** effect arises from magnetic susceptibility variation (e.g. tissue-air inter-

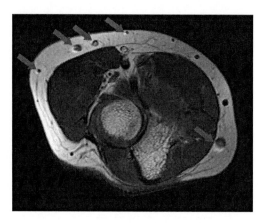

Figure 1.13: *Example of chemical shift artifacts of the first kind. This MR image shows a human shoulder. The chemical shift artifact is induced due to blood flow within the vessels.*

faces) or B_0-field inhomogeneities. Eddy currents can also distort EPI images and lead to signal loss. $\mathbf{T_2^*}$ **decay** during the formation of gradient echo train causes blurring.

1.2 Thermal therapy and Cryotherapy

In this study, liver metastases and tumors are a main topic, because 70% of all malignoms cause liver metastases. These liver metastases / tumors are often the reason for patient fatality, because only one third of all patients can be treated by surgical resection. In this case, alternative methods like minimal-invasive local thermotherapy have to be found. Such thermal therapies like Radio-Frequency Ablation (RFA) (Section 1.2.2), Laser Induced Thermal Therapy (LITT) (Section 1.2.1), High Intensity Focused Ultrasound (HIFU) (Section 1.2.3), Microwave Ablation [92], or Cryotherapy (Cryo) (Section 1.2.4) have to be monitored. MRI is the only imaging modality which allows for online non-invasive, spatially resolved, in-vivo temperature monitoring.

The advantages of minimally invasive therapies (LITT, RFA, Cryo) are the minor burden for patients, the benefit in quality of life, and the arbitrary repeatability. Nevertheless, each thermal treatment modality has several advantages and disadvantages (Table 1.2).

Cryotherapy is less painful than RFA and LITT and can induce larger lesions, but could induce ischemia and may lead to bleeding during and after thawing.

HIFU is a non-invasive thermal therapy, which can induce lesions of arbitrary size by screening the lesion voxel by voxel [32,54,80]. A particular challenge is the ablation of areas behind bone-structures like the ribs or skull or the ablation of moving organs, because the focal point of the beam has to follow the target-motion. Currently, a lot of studies are being done concerning HIFU on the brain or the prostate [32,54,60,80,96,119].

For RFA and LITT, lesion size is limited. But, malignant tissue has to be destroyed completely including a safety border. Solutions for this include the use of multiple applicators or embolisation. Most thermal therapies in clinical practice are performed using RFA. Radio-frequency ablations have been tested in several studies [8,9,18,55,137,147] and are cheaper than LITT. Thermal therapies using a laser have also been well evaluated [79,97,107,111,118]. In contrast to the metallic RF-needle which disturbs the acquired GRE phase data, the laser-fibers induce no susceptibility artifact within the phase images.

Table 1.2: *Overview of several heat sources [51].*

Type of Ablation	Mechanism	Advantages	Disadvantages
Cryo-ablation	Freeze-thaw cycle.	Virtual absence of pain. Can create large lesions and is effective in treating tumors in multiple lobes. Ability to reversibly test the effectiveness on an ablation site.	Lesions are significantly affected by blood flow. High complication rate.
High intensity focused ultrasound (HIFU)	Transducer driven by sinusoidal signals in a continuous wave or quasi-continuous-wave mode to generate ultrasound.	Ability to focus on the area under treatment. Completely non-invasive Good depth of penetration with the ability to pass harmlessly through tissues. Ablation of arbitrary lesion size.	Cannot be directed through air-filled viscera such as the lung. If bone is in the pathway, it absorbs the heat. Target tracking has to be performed to move the hot spot of the HIFU beam with the target.
Radio-frequency	Resistive heating by RF current.	Simple system design, proven effectiveness and worldwide availability. The complication profile is acceptable. Ability to treat different tumor types.	Limited extent of induced necrosis. Ablation zones can not exceed 4 cm unless the ablation probe is repositioned for another ablation.
Micro-waves	Heating by propagating electro-magnetic waves.	High temperature availability. Capable of forming large lesions in the presence of blood perfusion.	Complications include pleural effusion, hemorrhage, and abscess.
Laser	Converts light to heat.	**Fully compatible with MRI.** Can deliver controlled low energy through a variety of fiber configurations to achieve thin, continuous lesions in and around defined structures.	Expensive and bulky system. Limited ablation zone. Tissue charring around the tip of the fiber.

1.2.1   Laser induced thermal therapy

The Laser induced thermal therapy (LITT) is generally performed using optical radiation in the near-infrared wavelength range (from about 700 - 2000 nm). When photons are launched into the tissue, one of these three events occur: scattering, absorption, or exit from the tissue. If a photon is absorbed, the energy from the photon is converted into inter- and intra-molecular energy and results in the generation of heat within the tissue. At the same time, the absorption by the tissue limits the size of the lesion created by the laser irradiation. A compromise between penetration and absorption has to be found. After initial absorption, the generated temperature spreads through the tissue and enlarges the lesion, dependent on the perfusion of the tissue. Large vessels remove heat and reduce effective maximum temperatures.

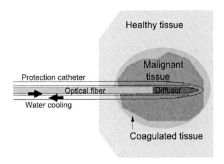

Figure 1.14: *Schematic illustration of laser induced thermal application.*

As heating continues and tissue temperature is elevated, several processes occur which lead to the destruction or death of the tissue:

- $T \geq 100\,°C$: Water in the tissue and intracellular compartments may vaporize and lead to rupture or explosion of cells or tissue components.

- $T \geq 60\,°C$: The proteins and cellular components of the tissue become severely denatured and coagulate leading to cell and tissue death.

- $T \geq 45\,°C$: Prolonged exposure leads to thermal denaturation of non-stabilized proteins such as enzymes. Cell death is not immediate. Longer heat exposure leads to the destruction of critical enzymes and cell death.

Optical absorption in the near-infrared range is generally due to fundamental molecular stretches. For wavelengths near 1000 nm, water is a primary absorber of optical energy. Applicators have diffuse characteristics and high energy deposition in the target-region. The induced effect of the laser depends on the wavelength, the exposure time, and the power density of the laser.

1.2.2   Radio-frequency ablation

A Radio Frequency Ablation (RFA) induces coagulation using an electromagnetic energy source with frequencies less than 900 kHz. Most devices operate in the range of 375-500 kHz. RF devices are divided into monopolar and bipolar ones. Monopolar RF-applicators have a single active electrode, whose current is dissipated at a return grounding pad. Bipolar devices have two active electrode applicators, which are usually placed in proximity to each other in order to achieve continuous coagulation between the two electrodes. A further type of RF applicators are the multiple expanded electrodes. Such RF applicators are built as an array of multiple electrode tines that expand from a single centrally positioned larger needle cannula.

1.2.3   High intensity focused ultrasound

Ultrasound wave emission is based on transducer vibration, typically in the range of 1 to 10 MHz. This results in dilatation and contraction modification of acoustical pressure. The acoustical pressure creates tissue movement (dilation and contraction). The movement amplitude is directly related to the pressure level. As the tissue response is not ideally elastic, energy is lost and converted into heat.

By using a spherical shaped transducer, the ultrasound beam is concentrated on the transducers focus point, resulting in a maximum pressure which is concentrated at this point. As tissue heating is directly related to maximum pressure, the necrotic lesion is formed at the transducer focus. The principle of HIFU therapy is the generation of a strongly focused ultrasound field by means of a special transducer. The induced focus is directed exactly to the region to be treated inside the human body.

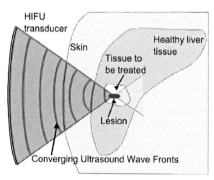

Figure 1.15: *Schematic illustration of HIFU.*

1.2.4 Cryotherapy

Cryotherapy is a method which destroys tissue by the application of low-temperature freezing. The freezing of tissue with rapid thawing leads to the disruption of cellular membranes and induces cell death. In the neck, chest, abdomen or pelvis, and extremities, cryo ablation is performed by using a closed cryo probe that is placed on or inside a tumor. In the main types of systems, argon gas and either gaseous or liquid nitrogen are used.

1.3 Temperature monitoring using MRI

Temperature monitoring can be used for several applications: hyperthermia ($40\,^{\circ}\mathrm{C} < T < 45\,^{\circ}\mathrm{C}$), hypothermia ($28\,^{\circ}\mathrm{C} < T < 35\,^{\circ}\mathrm{C}$) as well as thermal therapy ($T > 55\,^{\circ}\mathrm{C}$) and cryotherapy ($T < -40\,^{\circ}\mathrm{C}$). For these therapies, histological and macroscopic tests cannot be performed. Tissue status and temperature development have to be monitored using imaging modalities. Ultrasound, Computed Tomography (CT) or Magnetic Resonance Imaging (MRI) can be used to monitor in-situ-ablations. The preferred imaging modality is MRI. It can be used in five separate and distinct ways: planning, targeting, monitoring, controlling, and assessing treatment response. MRI provides a better soft part tissue contrast than CT or Ultrasound. CT induces radiation. Ultrasound provides a lower spatial resolution than MRI. Out of the three described imaging modalities, MRI is the only one that allows for temperature monitoring.

1.3.1 Overview of methods for temperature monitoring using MRI

For temperature monitoring, different physical parameters can be used like the equilibrium magnetization (M_0), T_1-contrast, diffusion-coefficient of water, and the chemical shift (Proton-Resonance-Frequency shift (PRF) and spectroscopy).

The **equilibrium magnetization** can be described by the Boltzmann distribution [1]:

$$M_0 = \frac{N\gamma^2\hbar^2 I(I+1)B_0}{3\mu_0 kT} = \chi_0 B_0 \tag{1.41}$$

where N is the number of spins per volume, γ is the gyromagnetic ratio, $\hbar$ is the Planck constant, I is the quantum number of the spin system ($1/2$ for protons), B_0 is the macroscopic magnetic field, μ_0 is the permeability of free space, κ is the Boltzmann constant, T is the absolute temperature of the sample, and χ_0 is the susceptibility. The relationship between χ_0 and temperature is known as the Curie law:

$$\chi_0 \propto \frac{1}{T} \tag{1.42}$$

Temperature changes based on equilibrium magnetization weighted images can be evaluated, because M_0 depends on the Boltzmann thermal equilibrium. The susceptibility which reflects the ratio of parallel and anti-parallel spin populations changes with temperature, not the equilibrium magnetization in the tissue itself. Between $37\,^{\circ}\mathrm{C}$ and $80\,^{\circ}\mathrm{C}$, the temperature sensitivity of M_0 is inversely proportional to the temperature and changes by $-0.30\pm0.01\,\%/^{\circ}\mathrm{C}$ [59]. It has to be considered that a temperature dependency of $-0.3\,\%/^{\circ}\mathrm{C}$ is small and requires a high SNR [13]. To eliminate effects from changes in T_1 relaxation time, long repetition times close to $10\,\mathrm{s}$ are required, making the method less useful for real-time applications [13,48].

Spectroscopy uses the distance between lipid- and water-peak within the spectra. It is the only MRI temperature monitoring method enabling absolute temperature monitoring. The distance between water and lipid peak shows a linear correlation with temperature and is used to calculate temperature [66,71,85]. Nevertheless, several challenges have to be overcome. The shielding of protons against the macroscopic magnetic field in water is temperature dependent, despite the fact that the shielding of lipids is independent of the temperature. The susceptibility of water protons is temperature independent; that of lipids is dependent on temperature. Both properties, susceptibility and shielding, induce a frequency shift and can lead to errors in temperature estimation, when water and lipids are in the same voxel, because both effects, susceptibility and shielding, cannot be separated. Additionally, multi-peak characteristics and availability of several types of fat make the use of the spectroscopy technique challenging. Furthermore, the temporal and spatial resolution of MR spectroscopy is not adequate for temperature monitoring (spatial resolution of 3-4 mm [69]; conditions discussed in Chapter 2.1).

Diffusion is characterized by the Brownian molecular movement. Increasing the temperature results in an increase of energy and an increase of the distance covered per time-unit. The relationship between the temperature and the diffusion coefficient can be written as [73]:

$$D \approx e^{\frac{-E_A(D)}{kT}} \tag{1.43}$$

where $E_A(D)$ is the activation energy of the molecular diffusion of water, κ the Boltzmann constant, and T the absolute temperature. The temperature dependence is described by:

$$\frac{dD}{D \cdot dT} = \frac{E_A(D)}{kT^2} \tag{1.44}$$

with a temperature sensitivity of about 2 %/°C. Random Brownian motion of molecules results in a Gaussian distribution of displacements. From diffusion coefficients D and D_{Ref}, acquired at two different temperatures T and T_{Ref}, respectively, the temperature change ΔT can be obtained:

$$\Delta T = T - T_{\text{Ref}} = \frac{kT_{\text{Ref}}^2}{E_A(D)} \left(\frac{D - D_{\text{Ref}}}{D_{\text{Ref}}} \right) \tag{1.45}$$

It is assumed that the temperature change is small ($\Delta T << T_{\text{Ref}}$) and that E_A is independent of temperature.

The diffusion method has been used to non-invasively measure temperature in vivo [6]. Its temperature sensitivity is high, but acquisition times are relatively long and in vivo implementations suffer from an extremely high sensitivity to motion. Single-shot echo planar imaging [6] and line-scanning techniques [93] have been used to reduce the acquisition time and motion sensitivity of this method. An additional in vivo complication occurs when the temperature dependence of the diffusion coefficient becomes nonlinear due to changes in the tissue condition. The mobility of water in tissue depends on barriers such as cellular structures, proteins, and membranes. Heat induced changes like protein coagulation can lead to large changes in the diffusion coefficient. In addition, nonlethal physiological effects such as ischemia in the brain [94] can also lead to large changes in diffusion coefficient.

In tissues with anisotropic diffusion, e.g. muscle fibers, the mobility of water protons is dependent on direction. For accurate temperature measurements, calculations of the full diffusion tensor, which describes the anisotropy, or the trace, which is rotationally invariant, might be necessary, but these methods require more acquisition time than a diffusion measurement in a single direction.

Lipid suppression is necessary in tissues containing fat, because fat has a different change of the diffusion coefficient with temperature.

A temperature gradient within a single voxel can cause phase dispersion due to PRF changes that may decrease the signal independent from D. The use of spin-echo methods is recommended to measure the diffusion coefficient.

In Summary, temperature accuracy and precision using the diffusion coefficient is not as good as needed for online monitoring during an intervention.

T_1 **relaxation** time can also be used to monitor temperatures non-invasively [7,102]. T_1 varies with temperature in the following manner [103]:

$$T_1(T) = T_1(T_{\text{Ref}}) \cdot e^{-\frac{E_A}{\kappa T}} \tag{1.46}$$

E_A is the activation energy of the relaxation process, κ the Boltzmann coefficient, and $T_1(T_{\text{Ref}})$ the T_1 relaxation time for the reference temperature T_{Ref}. One disadvantage of this kind of temperature monitoring using MRI is the tissue dependent correlation between temperature (T) and T_1. Tissue dependency can be explained by the variation of E_A for different tissues, dependent on the ratio between free and linked water molecules. The temperature dependence of the longitudinal relaxation times can also be described as:

$$T_1(T) = T_1(T_{\text{Ref}}) + m(T - T_{\text{Ref}}) \tag{1.47}$$

where $m = dT_1/dT$ is determined empirically for each tissue [21]. Temperature dependence was found to be in the order of $1\,\%/°C$ [75], with values of $1.4\,\%/°C$ in bovine muscle [19], $1-2\,\%/°C$ in liver [86], and $0.97\,\%/°C$ in fat [57].

The signal for both, spin echo or gradient echo images, can be expressed in terms of M_0, the flip angle α, the relaxation time T_1 and the repetition time TR as:

$$S = M_0 \cdot sin(\alpha)\frac{1 - E}{1 - cos(\alpha) \cdot E} \tag{1.48}$$

$$E = exp\left[\frac{-TR}{T_1(T_{\text{Ref}} + m(T - T_{\text{Ref}}))}\right] \tag{1.49}$$

The relative temperature sensitivity of the magnitude image $dS/(S \cdot dT)$ is related to the rate of signal change with relaxation dS/dT_1:

$$\frac{dS}{dT} = m\frac{dS}{dT_1} - \frac{S}{T} \tag{1.50}$$

The signal decreases with increasing temperature. Both, M_0 and T_1, change with temperature. The relaxation time increases and the magnetization decreases with increasing temperature. The second term on the right hand side represents the decrease in the equilibrium magnetization with increasing absolute temperature.

Using equations 1.48 and 1.50, the temperature sensitivity $dS/S \cdot dT$ is given by:

$$\frac{dS}{S \cdot dT} = \frac{m \cdot TR \cdot (1 - cos(\alpha)) \cdot E}{T_1(T_{Ref})^2 \cdot (1 - E) \cdot (1 - cos\alpha \cdot E)} - \frac{1}{T_{Ref}} \qquad (1.51)$$

Accuracy and precision of T_1-based thermal mapping depends on the accuracy of measuring and extracting T_1. Many accurate T_1 mapping methods such as inversion recovery and saturation recovery are very time consuming and not useful for monitoring thermal therapy, although single-shot methods partially alleviate the problem. Lipid suppression should be used, because lipids have a different T_1 change with temperature. Therefore, the presence of lipids within the monitored tissue is a potential source of artifacts. In addition, a temperature gradient within a single voxel can cause phase dispersion. Phase dispersion is induced by PRF changes, which decrease the received signal independent from T_1-changes. The use of spin-echo methods, which refocus the phase dispersion, eliminates this problem.

The quantification of temperature changes using T_1 effects is difficult. The temperature coefficient of the individual tissues is usually not known. Furthermore, the physiologic response of the living tissue to heat can seriously affect the quantification [144]. Nonlinear effects can occur if the tissue properties change, e.g. due to coagulation. In ex vivo tissue, coagulation has been found at temperatures as low as $43\,^{\circ}C$ [99]. Due to these problems, T_1-changes are often only used to perform a qualitative measurement of temperature distribution. If only a qualitative temperature measurement is needed, T_1 weighted images can be acquired rapidly and compared to or subtracted from baseline images acquired before heating. T_1 and the T_1-change per $^{\circ}C$ [10] increase with increasing field strength, but T_1 contrast diminishes [144]. Apart from SNR advantages at a higher field, T_1-based temperature mapping, using T_1-weighted images, appears more sensitive at a low field. Young et al. [144] described T_1-weighted thermo-sensitive gradient echo images to monitor LITT.

In clinical practice, the **PRF method** is the preferred MRI temperature monitoring method. One advantage is the tissue independence, despite fat and the strong linearity of the Larmor frequency shift and temperature difference. Fat suppression or water excitation reduces the influence of the fat signal and improves temperature accuracy and precision. A detailed description of the PRF-method can be found in the next section 1.3.2.

1.3.2 Proton Resonance Frequency shift (PRF) method

1.3.2.1 Dependency between Larmor frequency and temperature

The first observations were reported by Hindman in 1966 [53]. The resonance frequency of a nucleus in a molecule is determined by the local magnetic field:

$$B_{loc} = (1 - \delta)B_0 \tag{1.52}$$

where δ is called the shielding constant or shielding constant and is dependent on the chemical environment. As a result of nuclear shielding, the Larmor frequency ($\equiv$ resonance frequency) becomes:

$$\omega = \gamma B_0(1 - \delta) \tag{1.53}$$

More details on chemical shift and nuclear shielding can be found in [58].

In water molecules, the hydrogen nuclei ^1H are shielded from the macroscopic field by the electrons of the molecule. An ^1H nucleus in a free H_2O molecule is shielded more efficiently by the electron cloud than a nucleus in a H_2O molecule which is hydrogen bonded to another molecule. Hydrogen bonds between neighboring molecules distort the electronic configuration of the molecules, which reduces the electronic shielding. The fraction and the nature of the hydrogen bonds in water vary with temperature. As the temperature increases, the hydrogen bonds stretch, bend [53], and break [126]. On average, the H_2O molecules spend less time in a hydrogen-bonded state. Consequently, there is more electron shielding of the ^1H nucleus, and thus a lower local magnetic field B_{loc} and a lower proton resonance frequency. Because of the physical origins of hydrogen bonding among water molecules, electron shielding is considered to be a microscopic effect. A detailed description of these processes is given in [53] and [126]. Temperature measurement using the PRF is enabled by the temperature dependent variation of hydrogen-bridge-linkages. These hydrogen-bridge-linkages affect the electron-configuration in water molecules and lead to temperature dependent shielding δ of the macroscopic magnetic field B_0 for the hydrogen nucleus. This effect decreases with increasing temperature T and leads to a linear chemical shift of the resonance frequency:

$$\delta(T) = \delta_0 + \frac{d\delta}{dT} \cdot T \tag{1.54}$$
$$= \delta_0 + a \cdot T \tag{1.55}$$
$$a \approx -0.01 \, \mathrm{ppm/^\circ C} \tag{1.56}$$

a is the temperature coefficient for the PRF method. The average electron-shielding constant of pure H_2O varies approximately linearly with temperature by about $(-1.03 \pm 0.02) \cdot 10^{-8}$ per $^\circ$C over a wide range in temperatures from -15 $^\circ$C to 100 $^\circ$C [53], including the temperature range of interest for interventional procedures. Nevertheless, the tissue type independence of the PRF shift is only true for aqueous tissues. In water, the dependence of the PRF on temperature is attributed to changes in the hydrogen bonds, which are absent in fat [53].

Therefore, the temperature dependence in adipose tissue is almost completely determined by susceptibility effects (χ). The resulting temperature sensitivity of fat is some orders of magnitude smaller, indicating that thermometry inside fatty tissue is difficult.

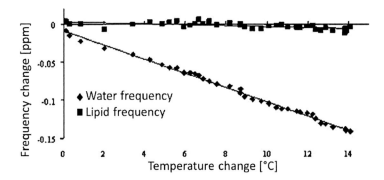

Figure 1.16: *Frequency change plotted over temperature change for lipid and for water [131]*

As mentioned above, the local magnetic field of hydrogen protons (B_{Proton}) depends on several parameters and equation 1.52 can only be used as approximation. The total shielding constant δ_t can be described as a sum of several terms:

$$\delta_t(T) = \delta_0 + \delta_\mathrm{d} + \delta_\mathrm{a} + \delta_\mathrm{W} + \delta_\mathrm{P} \qquad (1.57)$$

- δ_0 = inter-molecular shielding constant for the nucleus in the isolated molecule

- δ_d = main part of diamagnetic susceptibility

- δ_a = part, if molecular susceptibility of neighbor molecules is anisotropic

- δ_W = van der Waals-term

- δ_P = polarization term

The PRF varies linearly with the amplitude of the macroscopic magnetic field (B_{Proton}), the essential field at the paramagnetic nucleus of the molecule. The macroscopic magnetic field (B_0) does not include the effect of the microscopic currents, which additionally shields the proton. This can be divided into two parts:

- The microscopic shielding δ_{mi}, which is the shielding by the molecule induced by currents within the molecule itself.

- The macroscopic shielding δ_{ma}, which describes the shielding from the macroscopic magnetic field (B_0).

$$B_{\text{Proton}}(T) = B_0 \cdot (1 - (\delta_{mi}(T) + \delta_{ma}(T))) \tag{1.58}$$

$$= B_0 \cdot \left(1 - \delta_{mi}(T) - \frac{2}{3}\chi(T)\right) \tag{1.59}$$

Higher order terms of the microscopic shielding δ_{mi} and the susceptibility χ are neglected. For the temperature range of interest, the temperature dependence of both χ and δ_{mi} can be approximated as linear. The susceptibility change with temperature is 0.0026 ppm/°C in pure water and 0.0016 ppm/°C for muscle tissue in the temperature range of 30 °C to 45 °C [23]. Whereas the temperature dependence of the chemical shift is nearly constant for all tissue types (with the exception of adipose tissue), the temperature dependence of the susceptibility is tissue-type dependent [144]. For pure water and tissues with high water content such as muscle tissue, the temperature dependence of the shielding constant is much larger than that of the susceptibility. The temperature dependence of the susceptibility constant only has a small effect on thermometry applications in these tissues and errors remain within 10% of the temperature variation [23]. Therefore, most implementations of PRF thermometry in aqueous tissues assume only temperature effects of the shielding constant. More details about the effect of temperature dependent susceptibility in the PRF method can be found in De Poorter et al. [23].

If water and fat are both in one voxel it cannot be evaluated whether a phase change is induced by susceptibility change of fat or by PRF shift of water. In most organs, fat saturation or water excitation can be used for adequate temperature monitoring and for the reduction of lipid-artifacts.

1.3.2.2   Temperature monitoring using the standard reference PRF method

All GRE sequences can be used for temperature monitoring using the PRF method [58]. A temperature difference ΔT induces a frequency modulation of the MR-signal $\nu = a \cdot \Delta T$ and a temperature dependent phase change. The use of a reference measurement eliminates δ_0 (for example magnetic field inhomogeneities). The PRF method calculates relative temperatures using phase difference $(\Delta\varphi)$ images:

$$\Delta T = \frac{\Delta\varphi}{a \cdot \gamma \cdot B_0 \cdot TE} \tag{1.60}$$

$$= \frac{\varphi_T - \varphi_{\text{Ref}}}{a \cdot \gamma \cdot B_0 \cdot TE} \tag{1.61}$$

where φ_{Ref} is the reference phase and φ_T is the phase including the temperature dependent phase shift, B_0 the magnetic field strength, TE the echo time, γ the gyromagnetic ratio, and a the temperature coefficient.

The echo time TE can be optimized to increase the phase contrast-to-noise ratio (CNR). Optimizing the CNR increases temperature accuracy and standard deviation in the temperature image σ_T. In a GRE sequence, the temperature-dependent phase difference signal-to-noise ratio (SNR), $\text{SNR}_{\Delta\varphi}$, is estimated as follows:

$$\text{SNR}_{\Delta\varphi} \;=\; \frac{|\Delta\varphi(\Delta T)|}{\sigma_{\Delta\varphi}} \tag{1.62}$$

where $\Delta\varphi(\Delta T)$ is the phase difference and $\sigma_{\Delta\varphi}$ is the standard deviation of the phase difference image. With $\sigma_{\Delta\varphi} = \sigma/A$, where A is the signal amplitude, the phase difference SNR is directly proportional to the signal intensity:

$$\text{SNR}_{\Delta\varphi} \;\propto\; |\Delta\varphi(\Delta T)| \cdot A \tag{1.63}$$

The GRE signal amplitude A is dependent on the tissue parameters, ρ (tissue density), T_1 and $T_2^\star$, as well as the sequence parameters (TE, TR, and flip angle) of the GRE sequence. Assuming the tissue parameters are relatively constant, the signal intensity only depends on the sequence parameters. The GRE signal decreases exponentially with increasing TE and with the time constant $T_2^\star$, which accounts for the transverse relaxation and dephasing of the magnetization vector. Furthermore, the phase shift increases linearly with TE. The $\text{SNR}_{\Delta\varphi}$ dependence on the echo time can be written as:

$$\text{SNR}_{\Delta\varphi} \;\propto\; TE \cdot e^{-\frac{TE}{T_2^\star}} \tag{1.64}$$

Differentiating equation 1.64 with respect to TE yields the optimal TE in the temperature dependent phase imaging at TE $= T_2^\star$ [20, 69].

The standard reference „PRF method needs temporal stability of the external magnetic...“[1] field. „A drift of the external magnetic field, which can be caused for example by intense gradient utilization [38], results in an additional phase shift commonly referred to as phase drift. This phase drift causes incorrect temperature estimation during the thermal therapy. However, the phase drift can be measured with a reference phantom that remains at a fixed temperature if the external field drift is uniform over the image...“[1] [24, 25].

„Motion is the most prevalent problem for temperature monitoring using standard reference PRF phase mapping...“[2] „...For temperature monitoring during thermal treatment, motion artifacts can be divided into two categories, intra-scan motion and inter-scan motion, based on the time scale of the motion with respect to the image acquisition time.

---

[1] Viola Rieke, Kim Butts Pauly. MR Thermometry. J Magn Reson Imaging. 2008 February; 27(2): 376390. page 11

[2] Viola Rieke, Kim Butts Pauly. MR Thermometry. J Magn Reson Imaging. 2008 February; 27(2): 376−390. page 13 and 14

Intra-scan motion is caused by the movement of an object during MR image acquisition, resulting in a poor quality image with typical blurring and ghosting artifacts. These motion artifacts are not specific to PRF temperature imaging and can be reduced by accelerating the image acquisition...“[2] „...Trade-offs between acquisition time and SNR and temperature uncertainty have to be considered.

Inter-scan motion occurs due to motion or displacement of an object between the acquisition of consecutive images. As discussed earlier, temperature images obtained using the PRF method are usually reconstructed by calculating the phase difference between a reference background image acquired prior to heating and the current heated image. If motion is present between the acquisition of the images, the images are not registered to the baseline and artifacts in the temperature maps occur. New baseline images cannot be acquired once the thermal procedure has been started until the heated region has returned to baseline temperature. Unfortunately, many of the target areas for thermal therapy are in the abdomen and influenced by motion and by deformation.

A major source of motion, especially in the upper abdominal organs, is respiration. Respiration not only displaces the organs but also changes the susceptibility field. A calculation method for the magnetic field distribution due to an arbitrary distribution of bulk susceptibility has been given by Salomir et al. [122]. Even without tissue motion in the imaged region, lung filling can change the background phase enough to make the calculated temperature map useless. Because treatment durations in thermal therapy are in the order of several minutes, the treatments cannot be performed in a single breath-hold. Using multiple breath-holds is difficult, because reproducible breath-holding is hard to achieve.

But even without respiratory motion, displacement between images can occur. Thermal coagulation leads to structural changes and deformation of the treated tissue, which can be observed ex vivo without any other source of motion present. This heating-induced tissue motion is often not a simple global displacement [89]. The tissue swells in three dimensions, causing a local warping of the field distribution at the position of the swelling. In vivo, swelling during the treatment and changes in muscle tension, or peristalsis can also cause tissue displacement.“[2]

To reduce and solve the above discussed problems using the PRF-method, a reference-less temperature monitoring method was developed and evaluated in this thesis (Chapter 4).

---

[2] Viola Rieke, Kim Butts Pauly. MR Thermometry. J Magn Reson Imaging. 2008 February; 27(2): 376390. page 13 and 14

1.4   Dosimetry and heat transfer

1.4.1   Heat transfer and Bioheat transfer equation

After induction of heat with a heat source (like laser, radio-frequency source, or HIFU), temporal and spatial temperature distribution can be described by the so called BioHeat Transfer Equation (BHTE) [15,104]. The BHTE introduces spatially dependent tissue parameters and considers heat conduction, blood perfusion, thermo-regulatory reaction of the vessel system as well as metabolic changes. Spatial and temporal dependent temperature distribution after heat induction can be described by:

$$\rho(\vec{r}, t) \cdot c_p \cdot \frac{\partial T(\vec{r}, t)}{\partial t} - \nabla \lambda \cdot \nabla T(\vec{r}, t) = Q_L(\vec{r}, t) + Q_B(\vec{r}, t) + Q_M(\vec{r}, t) \tag{1.65}$$

where λ is the thermal conductivity coefficient, c_p is the specific heat capacity for constant pressure, and $\rho(\vec{r}, t)$ is the temporal and spatial dependent tissue density. The left term of equation 1.65 describes changes of the inner energy. Q_L characterizes the source term of the absorption energy, Q_B is the heat transmission, and Q_M describes metabolism, vaporisation, and condensation. Figure 1.17 shows a temperature distribution around a laser fiber simulated using equation 1.65.

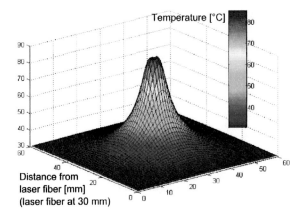

Figure 1.17: *Typical temperature distribution for heat transfer in tissue around heat source like a laser fiber.*

As an approximation, Q_B and Q_M are neglected in order to solve equation 1.65. The spherically symmetric solution-model for the heat distribution is displayed in figure 1.18. The temperature field around the sphere can be described as:

$$\frac{\partial T}{\partial t} = a\left(\frac{\partial^2 T}{\partial r^2} + \frac{2}{r}\cdot\frac{\partial T}{\partial r}\right) \tag{1.66}$$

$$a = \frac{\lambda}{c_p \cdot \rho} \tag{1.67}$$

where r characterizes the distance between an arbitrary point outside the sphere and the center of this sphere (Figure 1.18).

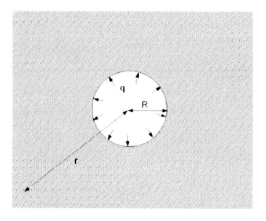

Figure 1.18: *Model for local heat source. Heat source of radius R and heat flux $\dot{q}$ in an infinity body.*

The boundary condition is a spatially constant heat flux $\dot{q}(t)$ through the surface of the sphere. If $R << r$, the sphere can be described as a punctual heat source, which has the following heat power:

$$\dot{Q}(t) = \frac{dQ_L}{dt} = 4\pi \cdot R^2 \cdot \dot{q}(t) \tag{1.68}$$

Initial conditions are:

$$T(r, t = 0) \quad : \quad T = 0 \tag{1.69}$$

$$r = R \quad : \quad \dot{q} = \frac{\dot{Q}_L}{4\pi \cdot R^2} = -\lambda \cdot \frac{\partial T}{\partial r} \tag{1.70}$$

$$r \to \infty \quad : \quad T = 0 \tag{1.71}$$

For $R \rightarrow 0$, $r = 0$, and $t = 0$ the Laplace transformation is:

$$T(r,t) = \frac{1}{(4\pi \cdot a)^{3/2} \cdot c_p \cdot \rho} \int_0^t \dot{Q}_L(\tau) \frac{exp\left(\frac{-r^2}{4a \cdot (t-\tau)}\right)}{(t-\tau)^{3/2}} d\tau \tag{1.72}$$

Two specific cases for equation 1.72 are discussed in the following section. First, a heat source of constant power $\dot{Q}_L(t) = \dot{Q}_c$ is examined. If the power of the heat source is constant, equation 1.72 results in an error-function:

$$T(r,t) = \frac{\dot{Q}_c}{(4\pi \cdot \lambda \cdot r)} \cdot erf\left\{\frac{r}{\sqrt{4a \cdot t}}\right\} \tag{1.73}$$

The property $lim_{x \rightarrow 0} erf(x) = 1$ of the error-function can be used for $t \rightarrow \infty$ and simplifies equation 1.73:

$$T(r) = \frac{\dot{Q}_c}{(4\pi \cdot \lambda \cdot r)} \tag{1.74}$$

Temperature distribution in equation 1.74 is temporally independent and only depends on the distance from the heat source; it describes a steady state.

Second, an infinity short power pulse at $r = 0$ for $t = 0$ using energy Q is applied:

$$T(r,t) = \frac{Q}{(4\pi \cdot a \cdot t)^{2/3} \cdot c_p \cdot \rho} \cdot exp\left(\frac{-r^2}{4a \cdot t}\right) \tag{1.75}$$

In equation 1.75, temperature behavior is $T \rightarrow \infty$ for $t = 0$ and $r \rightarrow 0$. For a constant position $r = const$, temperature varies with time. The temperature increases from $T = 0$, has a maximum at $T_{max} = r^2/6a$, and decreases back to $T = 0$ (Figure 1.19). Equation 1.72 is true for $R \rightarrow 0$, and works fine for small radii $R > 0$ if $R << r$. The temperature field can be described using a convolution of the weighting function $(w(r,t))$ and the heat power $(\dot{Q}(t))$:

$$T(r,t) = \int_{-\infty}^{+\infty} \dot{Q}(\tau) \cdot w(r, t-\tau) d\tau \tag{1.76}$$

The weighting function can be defined using equation 1.75:

$$w(r,t) = \frac{Q}{(4\pi \cdot a \cdot t)^{2/3} \cdot c_p \cdot \rho} exp\left(\frac{-r^2}{4a \cdot t}\right) \tag{1.77}$$

1.4.2 Arrhenius damage integral and peak temperature

Temperatures above $45\,°C$ lead to protein denaturation (see Table 1.3), the breaking of disulfide-boundaries and the destruction of molecules. **Arrhenius** published the following equation to describe the dependency between the temperature T and the rate of chemical reactions k [3, 117]:

$$k = A \cdot exp\left(\frac{-E}{R \cdot T}\right) \tag{1.78}$$

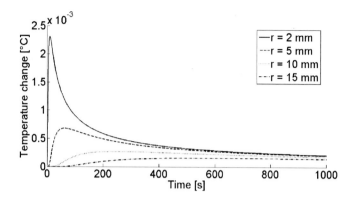

Figure 1.19: *Temperature distributions for an infinitely short heat pulse for certain distances r from the heat source. The shown graphs are calculated using equation 1.75 and following parameter:* $a = 1.50515 \cdot 10^{-7}\, m^2/s$, $c_p = 3590\, kg/K$, $\rho = 1050\, kg/m^3$ *and,* $Q = 12\, J/s$.

where A and E are the Arrhenius parameters and R the gas constant ($R = 8.3145\ \frac{J}{K \cdot mol}$). The parameter E characterizes the activation energy of the reaction and A the action entropy. A quite high activation energy is necessary for protein denaturation (Figure 1.20).

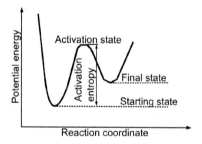

Figure 1.20: *Certain energy states in a chemical reaction.*

Nevertheless, coagulation can appear at low temperatures ($T \approx 60\,°C$, see Table 1.3). When molecule boundaries are broken, tertiary structure is lost and degrees of freedom are generated. Increasing degrees of freedom leads to an increase of entropy, which can be used to compensate for a high activation energy. This is also the reason for the irreversible character of the coagulation process.

If c_n is the concentration of native proteins and Δc_n is the denatured part after a time interval

Table 1.3: *Temperature induced effects on biological tissues [50].*

Temperature Range [°C]	Time Requirements	Physical Effects	Biological Effects
< −50	>10 min	Freezing	Complete cellular destruction
0−25		Decreased permeability	Decreased blood perfusion, decreased cellular metabolism, hypothermic killing
30−39	No time limit	No change	Growth
40−45	30−60 min	Changes in the optical properties of tissue	Increased perfusion, thermo-tolerance induction, hyperthermic killing
45−50	>10 min	Necrosis, coagulation	Protein denaturation, unsubtle effects
>50	After ≈2 min	Necrosis, coagulation	Cell death
60−140	Seconds	Coagulation, ablation	Protein denaturation, membrane rupture, cell shrinkage
100−300	Seconds	Vaporization	Cell shrinkage and extracellular steam vacuole
>300	Fraction of a second	Carbonization, smoke generation	Carbonization

Δt, then we can postulate the following equation:

$$\frac{\Delta c_n}{\Delta t} = -k \cdot c_n \qquad (1.79)$$

Temperature change during the denaturation process can lead to a time dependent reaction rate:

$$c_n(t) = c_n(0) \cdot exp(-kt) \Rightarrow kt = -ln\left[\frac{c_n(t)}{c_n(0)}\right] \qquad (1.80)$$

$k \cdot t$ can be interpreted as the damage integral $\Omega(t)$:

$$\Omega(t) = -ln\left[\frac{c_n(t)}{c_n(0)}\right] = \int_0^t A \cdot exp\left(-\frac{E}{RT}\right) dt \qquad (1.81)$$

Interpretation of the damage integral:

- $\Omega = 0$: all proteins are native; no protein coagulated

- $\Omega \to \infty$: all proteins are coagulated

- $\Omega \geq 1$: more than 63% of proteins are denatured and cell death can be expected.

A second model used to estimate tissue status is the **peak temperature model**, which disregards the temperature history of the subject. A lethal thermal damage is assumed to occur above a critical temperature. Below the threshold temperature nonlethal thermal damage is expected [145].

Chapter 2

Temperature accuracy and precision using standard reference PRF method on phantoms and non-moving organs

This chapter handles with temperature monitoring using MRI at non-moving organs like the brain. Temperature accuracy and precision are discussed.

2.1 Motivation, requirements, and applications

Temperature mapping can be used to monitor and control minimally invasive treatments such as focal brain thermo therapy [8,9,32,39,54,79,81,112,137–139,147], focal brain hyperthermia [62] and whole brain hypothermia [4,98,120,121,133,146]. In comparison to other imaging modalities, MRI alone allows for non-invasive, three-dimensional real-time temperature mapping.

Global cerebral hypothermia is often used to improve clinical outcomes after cerebral hypoxia or ischemia probably by decreasing the brain's energy and oxygen requirements [4,98,120,121,133,146]. Whether this is achieved either by cooling the whole body or by selectively cooling the head, it is important to keep the brain within a certain temperature range. Focal cerebral hyperthermia is clinically used in combination with radiation therapy or chemotherapy to augment clinical response. Focal hyperthermia is a low-temperature elevation therapy which requires well controlled temperatures in the narrow range between 43 °C and 45 °C.

Minimally invasive focal thermo therapy is used clinically to directly treat brain tumors. This involves the application of heat with effective peak temperatures ranging from 50 °C to 80 °C. As opposed to focal hyperthermia which is used to augment radiation or chemotherapy, focal thermo therapy involves temperatures that are high enough to kill neoplastic tissue primarily.

Focal heating can be implemented using a variety of heat sources (Section 1.2) including LITT
(Section 1.2.1), RFA (Section 1.2.2), and HIFU(Section 1.2.3).

It is especially advantageous that MRI can be used to verify and characterize malignant and
healthy tissue prior to, during, and after thermo therapy, and can also be used to monitor
temperature distribution during heating. The PRF method [11, 53, 58, 67, 87, 109, 142] is typically used for MR temperature mapping at field strengths higher than 1T [56] (Section 1.3.2).
For acquiring the phase images used for temperature evaluation based on the PRF method,
the gradient echo (GRE) sequence is used as gold standard. In many cases the standard GRE
sequence leads to relatively long acquisition times under optimized conditions for temperature
mapping (Section 1.1.5). As thermal therapy faces in many cases challenges, such as large
treated volumes and fast heating, techniques for temperature monitoring must be able to
provide both rapid measurement and high precision simultaneously [39, 81, 138, 139]. Critical
structures have to be spared from treatment to minimize adverse effects. Particularly in the
brain, heating of non-target areas has to be controlled in order to avoid damage to critical
structures. For example, during HIFU treatment of the brain, regions near the cranial bone
must be monitored, as ultrasound absorption of the bone can lead to heating of the cerebral
cortex [64, 88, 95]. In many cases a large volume, containing more than the treated area, should
be covered by MR temperature mapping in a sufficiently short time. There are a handful of
pulse sequences that can be used to provide PRF measurements, all of which are based on
variations of the GRE sequence (Section 1.1.6). These have been designed to decrease acquisition time to allow for real-time monitoring.

This study was conducted to compare the quality of temperature mapping of the ss EPI sequence to standard GRE and seg EPI sequences (Section 1.1.4) using the standard reference
PRF thermometry (Section 1.3.2.2). These were tested at both 1.5 T and 3 T on commercial
whole-body clinical scanners (MAGNETOM Avanto and MAGNETOM Trio (Tim system),
Siemens Healthcare, Erlangen, Germany). Parameters of the sequences were adapted to the
conditions in human brain.

2.2 Material and methods for evaluation of temperature monitoring of non-moving systems

2.2.1 Setup for temperature monitoring

Table 2.2.1 lists the acquisition parameters of the acquired sequences. All of the TRs listed
in Table 2.2.1 are calculated for the parallel acquisition of seven slices. The GRE sequence
collects one line of k-space in each of the 7 slices per read-out and is repeated 128 times.
The seg EPI sequence collects 13 lines of k-space for each of the 7 slices per read-out and is
repeated 10 times. The ss EPI sequence measures all lines of k-space in one slice per readout,
and is not repeated (Section 1.1.6).

Table 2.1: *Measurement parameters for temperature precision maps.*

	TE [ms]	TR [ms]	Average	TA for 7 slice [s]	α	BW [Hz per voxel]	GRAPPA factor
GRE	17	157	1	7,5	25	160	3
Seg EPI	17	254	1	3.3	45	500	-
Ss EPI	17	300	1	1.3	45	1562	2
	17	300	4	4.9	45	1562	2
	30	600	1	2.4	60	752	2
	30	600	4	6.6	60	752	2
	50	700	1	2.8	65	752	2
	50	700	4	7	65	752	2
	70	900	1	3.5	70	752	2
	70	900	4	7.7	70	752	2

Isotropic voxel of 2 mm³; TR= repetition time, TE=echo time, α=flip angle, TA=acquisition time, BW=Bandwidth, GRE=GRadient Echo, EPI=Echo-planar Imaging), GRAPPA (Generalized Autocalibrating Partially Parallel Acquisition (k-space based parallel imaging technique).

For selecting suitable sequence parameters, especially a suitable TR, several factors had to be taken into account. In order to acquire adequate temperature maps the sequence must:

- have a short sequence duration resulting in high temporal resolution in order to track rapid temperature changes

- provide the ability to monitor a large volume of tissue including both target and appropriate non-target tissue

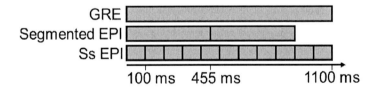

Figure 2.1: *Time scale of the three sequences used for temperature monitoring at volunteer brain.*

These factors are particularly important when using thermo therapy modalities, such as HIFU, which create rapid temperature changes, and which may affect sensitive non-target tissue. Hynynen demonstrated a temperature rise of 0.022 °C/W in 20 s with HIFU in the brain (600-1080 W) [95] and Mougenot one of 14 °C in 100 second with HIFU of rabbit leg muscle [88].

Essentially, both of these studies have demonstrated that HIFU induced temperature changes can be as rapid as 1 °C/s. Furthermore, HIFU therapy of brain tumors has been shown to induce heating of the brain surface due to skull heating [64]. This non-target heating requires that not only the target, but also the non-target areas, including the skull and the subjacent tissue, are monitored during the procedure using temperature mapping.

While short TR and TE allow for fast measurements, the PRF method is theoretically

Table 2.2: *Overview of $T_2^\star$ relaxation times in humane brain [31, 40, 70, 100].*

Tissue	$T_2^\star$ [ms]	
	1.5 T	3 T
gray matter	71.5	49.3
white matter	66.5	48.8

most sensitive to temperature measurements, if TE is in the same range as $T_2^\star$ [31, 40, 70, 100] (Section 1.3.2). $T_2^\star$ values for the brain are 48.8 ms / 49.3 ms for white / gray matter at 3 T and 66.5 ms / 71.5 ms for white / gray matter at 1.5 T [40] (Table 2.2). Therefore, a compromise between short measuring time (short TE), high resolution, and accuracy of the temperature scale (long TE) was necessary, and a TE of 17 ms was chosen for the GRE and seg EPI sequences (Table 2.2.1). Because the ss EPI sequence allows for very fast temperature mapping, measurements with longer TEs could be performed, which would provide higher temperature precision. For the ss EPI sequence, the TE was increased until the $T_2^\star$ relaxation time was reached, in order to maximize the SNR. At 3 T measurements were performed with TEs of 17 ms, 30 ms and 50 ms and at 1.5 T with TEs of 17 ms, 30 ms, 50 ms and 70 ms. In order to improve SNR and to keep the acquisition time short, the bandwidth was adapted for each sequence to be as low as possible (Table 2.2.1). A field of view (FOV) of 256 mm^2 with a symmetrical matrix size of 128 and a slice thickness of 2 mm was chosen for all sequences to have an isotropic resolution with a length of 2 mm^3. Parallel imaging was applied with a GRAPPA-factor of 3 for the GRE sequence and with one of 2 for the ss EPI sequence. In both cases, 24 reference lines were used for the calibration process in the reconstruction. For the seg EPI sequence an EPI factor (ETL) of 13 was used (Section 1.1.6).

Parameters were chosen as compromise between image quality, temperature map quality, short acquisition time, and a high resolution temperature scale.

In order to compare performance on both the 1.5 T and the 3 T systems, the parameters of the three pulse sequences were kept the same for 1.5 T and 3 T. As well, the TE was kept the same between the three sequences. For a sufficient high SNR, the phase precision is inversely proportional to the SNR [26]. This means that the minimal temperature standard deviation σ_T measured by MRI is restricted by:

$$\sigma_T = \frac{\sqrt{2}}{\text{SNR} \cdot B_0 \cdot \gamma \cdot \alpha \cdot \text{TE}} \tag{2.1}$$

To compensate for temporal instability of the main magnetic field, i.e. baseline phase drift, zero-order compensation was performed by subtracting the spatially averaged apparent temperature of non heated regions selected by the user [128]. For such a B_0 correction, a defined area inside the brain has to be chosen. The distance from the area being treated is chosen where: a) no heating and movement occurs during the intervention and b) the signal of the magnitude image is as intense and homogeneous as possible. The B_0 correction is carried out as a comparison of phase signals within this defined area of the present phase image and of the reference phase image.

For real-time monitoring of a calculated temperature during an intervention, a region of interest of variable size can be positioned in the magnitude image for visualization of temperatures in the treated region (equation 1.60).

Data was evaluated using MATLAB (Mathworks, Natick, MA, USA version 6.0). Difference-Phase-Maps (B_0-corrected difference between reference phase image and the present phase image) were imported into MATLAB and were processed according to equation 1.60 (Section 1.3.2.2).

Images with visible motion artifacts, caused, for example, by swallowing, sneezing or coughing, were omitted from this study because displacements caused by gross motion which occurs between the reference and the present acquisition will lead to an under- or overestimation of temperature. Images with such artifacts were omitted if the average phase difference between the reference and present phases in a predefined ROI within the brain was greater than a phase change equivalent to 5 °C in the unheated brain.

2.2.2 Comparative assessment of the ss EPI sequence accuracy

The GRE sequence has been tested extensively in the past and is seen as the gold standard for real-time MR temperature mapping [112]. Temperature measurement with the seg EPI used in this study was validated previously by Cernicanu et al. [109].

In order to determine the accuracy of the ss EPI sequence for temperature measurement, it was compared to the GRE sequence during three periods of heating of swine skeletal muscle using HIFU (80 W; 10 s). The spherical fixed focus single element HIFU transducer works with a frequency of 1.7 MHz and has a diameter of 60 mm. The HIFU transducer generates a focus in a distance of 60 mm with a width of 1.07 mm and a length of 8.9 mm. Temperature measurements with the two sequences GRE and ss EPI were performed sequentially in different locations within the same meat sample. For the experiment a homogeneous meat sample was chosen to have similar conditions for both experiments.

2.2.3 Evaluation of temperature precision

Nine healthy volunteers (age 25-43 years, four male, five female) participated in the study. Four were examined on a 1.5 T system (MAGNETOM Avanto, Siemens Healthcare, Erlangen, Germany) and five on a 3 T system (MAGNETOM Trio, Siemens Healthcare, Erlangen, Germany). Each volunteer underwent several measurements over approximately 12 minutes. For the GRE and seg EPI sequences six measurements were performed on the 3 T system and

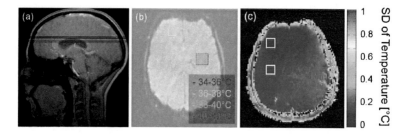

Figure 2.2: *(a) Illustrates the location of the 7 axial slices on a sagittal localizer of a volunteer's head. (b) Real-time image of a single axial slice shown in an online window as overlay of the temperature map on the magnitude image. MR image was measured with the ss EPI at 3 T with an axial slice thickness of 2 mm and a distance factor of 20%. (c) Standard Deviation (SD) map of temperature [°C] calculated in a post-processing step for a ss EPI sequence at 3 T, with the two exemplary square ROI's (10x10 voxel) used in this study for SNR, motion and SD estimation.*

four were performed on the 1.5 T system. For the ss EPI two measurements were performed for each combination of TE and averaging factor. All acquisitions were performed with a 12-channel phased array head coil (Siemens Healthcare, Erlangen). The positions of seven recorded axial slices of the head are displayed in Figure 2.2a on a sagittal image of the head. A typical real-time image of one of these axial slices is demonstrated in Figure 2.2b. Two regions of interest (ROI) can be seen, one for B_0 correction (yellow border; frontal lobe of the brain) and one for temperature visualization (red border) (Figure 2.2b). Both ROI's can be altered by the user both in size and position.

Analysis of measurements was performed by generating standard deviation maps from series of consecutive temperature measurements (Standard Deviation (SD) maps, Figure 2.3). Resulting SD maps from this post-processing provide illustrations of standard deviation values of all voxels in a slice over 12 minutes (Figure 2.2c) and allow for an estimation of the temperature precision of the three pulse sequences. For the evaluation of the temperature precision, a Box-Whisker plot of the SD calculated for both ROI's shown in Figure 2.2c was generated. A Box-Whisker plot visualizes both the calculated medians as well as the 10, 25, 75, and 90% margins. Such a Box-Whisker plot is a convenient way of graphically depicting groups of numerical data through their five-number summaries: the smallest observation, the lower quartile (25%), the median, the upper quartile (75%), and the largest observation. It may also indicate which observations, if any, might be considered outliers. The calculations take into account both ROI's (Figure 2.2c) of all 7 slices in the brains of all volunteers. The medians visualized in the Box-Whisker plots describing the temperature precision. ROI 1 was placed in the most challenging area of the brain, namely the frontal lobe adjacent to the frontal sinuses, where susceptibility artefacts would be strongest. ROI 2 was placed in the deep brain, adjacent to the lateral ventricles, in order to acquire more stable / ideal data.

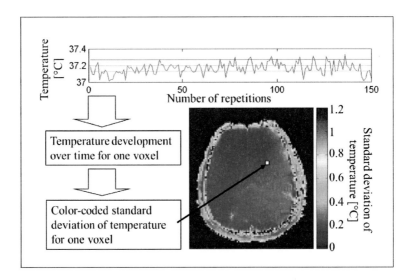

Figure 2.3: *Above an exemplary temperature distribution over time for one voxel in a volunteers unheated brain is shown. For each voxel in each slice a standard deviation of the temperature over a time duration of 12 min is calculated. In the lower right corner a map is displayed, which shows the calculated SD-values for each voxel as a characterization of the temperature precision.*

The SNR of the magnitude images was calculated using the dual acquisition method [33, 41]:

$$\text{SNR} = \sqrt{2} \left(\frac{S_1}{\text{SD}_{1-2}} \right) \tag{2.2}$$

where S_1 is the mean signal intensity in the ROI on the reference magnitude image in the brain (same homogeneous region as SD-value; Figure 2.2b), and SD_{1-2} is the standard deviation in the same ROI on a subtraction image. The subtraction image was calculated by subtracting the reference image from the first magnitude image image acquired during the intervention. The factor $\sqrt{2}$ arises because the noise is derived from the difference image. For each pulse sequence the SNR was averaged over the acquired 7 slices and over all volunteers examined at a given field strength (Table 2.4, Table 2.3).

The SNR of the phase difference images was also calculated using the dual acquisition, subtraction method. S_1 is the mean signal intensity in the ROI on the reference phase image in the brain (same homogeneous region as SD-value; Figure 2.2b), and SD_{1-2} is the standard deviation in the same ROI on a subtraction image. The subtraction image was calculated by subtracting the reference and the first temperature phase image. The SNR within the phase difference images was calculated for each repetitive acquired image of each volunteer and for all slices.

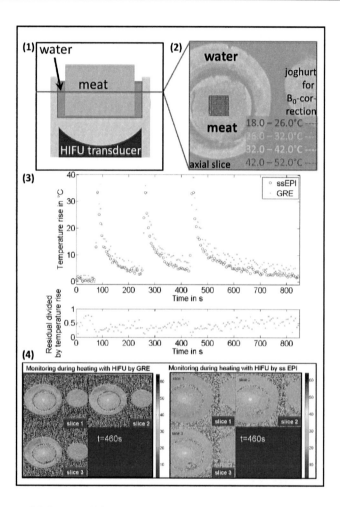

Figure 2.4: *(1) Drawing of the HIFU setup in the sagittal plane. The meat is placed in a water bath within the HIFU setup. The HIFU transducer is placed central under the meat. (2) Typical online temperature map visualization during heating in an axial plane. The image shows the area, which will be heated by the HIFU transducer. A yoghurt is placed next to the HIFU setup for a B_0-correction. (3) Temperature progression over time during heating using HIFU in swine skeletal muscle. Three heating periods are shown. The transducer fired with 80 W over a period of 10 s and a break of 3 min. The graph shows a comparison of temperature values acquired with the GRE and ss EPI sequences at a 1.5 Tesla system (MAGNETOM Avanto, Siemens Medical Solutions). (4) Consecutive spatial temperature maps [°C] were acquired for 3 slices. First, a measurement was performed using the GRE sequence and second, a measurement using the ss EPI sequence. Shown temperature maps are acquired for t=460 s.*

2.3   Accuracy and precision results

2.3.1   Comparative assessment of ss EPI sequence accuracy

The ex-vivo experiments performed in the swine model during heating showed a strong correlation between temperature measurements acquired using the GRE and ss EPI sequences as
exhibited on figure 2.4. The residual between the GRE and ss EPI divided by the temperature
rise is less than 1° C throughout the entire measurement. As shown in figure 2.4, a temperature increase of more than 30 °C occurs during heating with 80 W over a period of 10 s. The
experiment was performed twice. First, using the GRE sequence to monitor the temperature
distribution and second, using the ss EPI sequence. Despite the sequentially character of the
whole experiment (first monitoring with GRE, afterward with ss EPI) both graphs, the one of
the values acquired using the GRE sequence and the one using the ss EPI) match.

2.3.2   Evaluation of temperature precision

An example of SD maps from phase images of the brain of one volunteer is shown in Figure
2.5 for all three sequences at 3 T. All sequences were applied over the duration of 12 minutes.
The standard deviation (SD) was < 1°C in all brain areas for all three sequences. Figure 2.7
demonstrates the distribution of the SD in the two ROI's described in figure 2.2c.

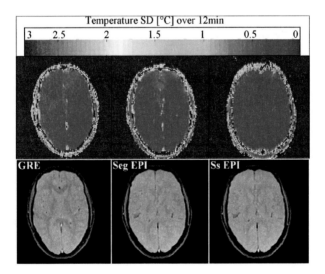

Figure 2.5: *Standard deviation maps of temperature data and associated magnitude images
in human head of a single volunteer measured with GRE, seg EPI and ss EPI on a 3 T
MAGNETOM Trio using a TE of 17 ms (no averaging).*

The overall median SD at 3 T across all volunteers and slices was $0.39 \pm 0.11°C$ for the seg EPI sequence and $0.37 \pm 0.11°C$ for the GRE sequence (Figure 2.7 and 2.6). At 1.5 T, median temperature precisions of $0.58 \pm 0.28/0.63 \pm 0.25°C$ were calculated for the GRE / seg EPI (Figure 2.7 and 2.8).

To compare the ss EPI to both the GRE and the seg EPI sequences, the temperature precision over 12 minutes was examined with the ss EPI sequence using both the same TE as the GRE and seg EPI sequences (17 ms), as well as longer TEs 30 ms, 50 ms, and 70 ms made possible by the fast nature of the ss EPI sequence. As well the ss EPI sequence was run with both a single average, and with an average factor of 4.

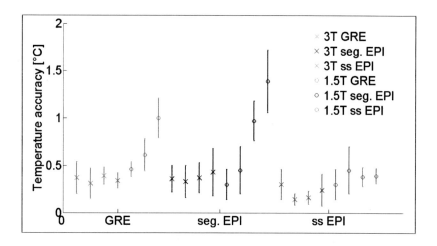

Figure 2.6: *Temperature accuracy reached for several volunteers at different field strengths using the GRE, seg EPI, and ss EPI sequences. The shown values are averages over 7 slices within the two ROI's shown in figure 2.2.*

At 3 T the median temperature precision over time was $0.52\pm0.08\,°C$ / $0.29\pm0.08\,°C$ / $0.16\pm0.06\,°C$ for a TE of 17 ms / 30 ms/ 50 ms for the ss EPI sequence. Using an average factor of 4 the temperature precision over time increased to $0.32\pm0.01\,°$ C / $0.21\pm0.03\,°$ C / $0.14\pm0.03\,°$ C for a TE of 17 ms / 30 ms / 50 ms. For average factors of $1/4$ the temperature precision over time increased from $0.52\,°$ C to $0.16\,°$ C / $0.32\,°$ C to $0.14\,°$ C as the TEs converged on the $T_2^\star$ relaxation time of the brain. Averaging had the largest effect on data acquired using the shorter TEs which is the least similar to the $T_2^\star$ relaxation time of the brain.

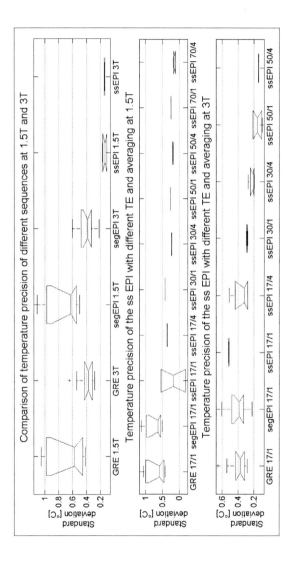

Figure 2.7: *Box-Whisker plot of the standard deviation of the temperature over 12 minutes for the three sequences (GRE, seg EPI, ss EPI), both field strength's (1.5T and 3T) and different TE and averaging factors for the ss EPI. Standard deviation was calculated within both ROI's in figure 2.2c. The box margins are given by the median (red) and the values of the SD that comprise 25% and 75% (blue) of all considered data points. The bar limits are defined by the extreme values that are exceeded by 10% and 90% (black) of all data points. The results contain all 7 slices from all volunteers. First row of the image shows an overview of the combinations of the two field strengths, 1.5T and 3T, and three sequences, GRE, segEPI, and segEPI. Row two and three show more detailed information to each tested sequence with the corresponding imaging parameter. The second graph (counted from the top) shows the detailed information at a 1.5T scanner and the third graph the information at a 3T scanner. The x-axis of the second and third graph is labeled in the following manner: sequence TE[ms] / averaging factor.*

Table 2.3: *Temperature precision for 3 T.*

$B_0{=}3\mathrm{T}$	Median temperature precision over time [°C]	Median temperature precision over space [°C]	Median SNR of magnitude image	Median SNR of phase difference image
GRE **TE=17ms** **1 Average**	0.37 (±0.11)	0.55 (±0.31)	63.4 (±3)	86.1 (±22)
Seg EPI **TE=17ms** **1 Average**	0.39 (±0.11)	0.48 (±0.39)	68.3 (±3)	76.2 (±28)
Ss EPI **TE=17ms** **1 Average**	0.52 (±0.08)	0.42 (±0.07)	65.2 (±7)	89.6 (±15)
Ss EPI **TE=17ms** **4 Averages**	0.32 (±0.01)	0.38 (±0.02)	70.3 (±4)	98.8 (±5)
Ss EPI **TE=30ms** **1 Average**	0.29 (±0.08)	0.24 (±0.08)	104.6 (±9)	154.8 (±11)
Ss EPI **TE=30ms** **4 Averages**	0.21 (±0.03)	0.19 (±0.01)	140.7 (±5)	198.9 (±6)
Ss EPI **TE=50ms** **1 Average**	0.16 (±0.06)	0.16 (±0.05)	148.6 (±14)	231.4 (±27)
Ss EPI **TE=50ms** **4 Averages**	0.14 (±0.03)	0.14 (±0.02)	174.3 (±6)	255.9 (±9)

B_0=field strength, GRE=Gradient Echo, seg EPI= Segmented Echo planar Imaging, ss EPI= Single Shot Echo planar Imaging.

At 1.5 T the median temperature precision over time was 0.54±0.02° C / 0.41±0.02° C / 0.27±0.04° C / 0.25±0.03° C for a TE of 17 ms / 30 ms / 50 ms / 70 ms for the ss EPI sequence. Using an average factor of 4 the temperature precision over time at 1.5T was 0.38±0.01° C / 0.23±0.01° C / 0.21±0.01° C / 0.19±0.03° C for a TE of 17 ms / 30 ms / 50 ms / 70 ms. For an average factor of 1/4 the temperature precision over time increased from 0.54° C to 0.25° C / 0.38° C to 0.19° C as the TEs converged on the T_2^* relaxation time of the brain.

Table 2.4: *Temperature precision for 1.5 T.*

$B_0{=}\mathbf{1.5T}$	Median temperature precision over time [°C]	Median temperature precision over space [°C]	Median SNR of magnitude image	Median SNR of phase difference image
GRE **TE=17ms** **1 Average**	0.58 (±0.28)	0.55 (±0.31)	54.2 (±3)	67.3 (±8)
Seg EPI **TE=17ms** **1 Average**	0.63 (±0.25)	0.48 (±0.39)	50 (±3)	62.4 (±13)
Ss EPI **TE=17ms** **1 Average**	0.54 (±0.02)	0.54 (±0.02)	57.1 (±7)	68.1 (±5)
Ss EPI **TE=17ms** **4 Averages**	0.38 (±0.01)	0.41 (±0.01)	66.6 (±3)	96.1 (±2)
Ss EPI **TE=30ms** **1 Average**	0.41 (±0.02)	0.45 (±0.02)	68.9 (±4)	82.5 (±3)
Ss EPI **TE=30ms** **4 Averages**	0.23 (±0.01)	0.34 (±0.01)	85.3 (±2)	111.3 (±2)
Ss EPI **TE=50ms** **1 Average**	0.27 (±0.04)	0.32 (±0.05)	87.3 (±5)	117.3 (±19)
Ss EPI **TE=50ms** **4 Averages**	0.21 (±0.01)	0.23 (±0.02)	110.1 (±3)	161.0 (±1)
Ss EPI **TE=70ms** **1 Average**	0.25 (±0.03)	0.28 (±0.05)	94.4 (±3)	134.0 (±16)
Ss EPI **TE=70ms** **4 Averages**	0.19 (±0.03)	0.21 (±0.05)	173.0 (±13)	188.6 (±24)

B_0=*field strength, GRE=Gradient Echo, seg EPI= Segmented Echo planar Imaging, ss EPI= Single Shot Echo planar Imaging.*

The standard deviation of temperature over space within one phase-difference image was also calculated (Tables 2.4 and 2.3). For all three sequences, the standard deviation of the temperature over time and standard deviation of the temperature over space are in the same range for each TE and each average factor tested. The feasible standard deviation over both time and space of the ss EPI sequence is in the same range as both seg EPI and GRE sequences when the same TE (17 ms) and average factor (one) were used (Figure 2.7). For the ss EPI sequence, increasing the TE up to the T_2^* relaxation time of the brain increased the temperature precision at both 1.5 T and 3 T. As well, averaging increased the temperature precision of ss EPI over time and over space, particularly for the smaller TE. Averaging was found to be unnecessary when using a TE $\approx$ T$_2^*$ of brain.

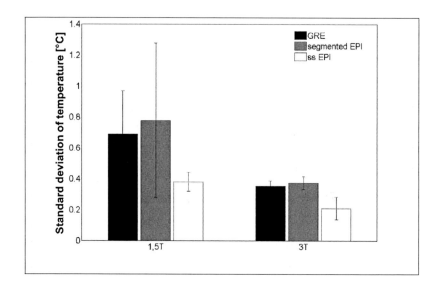

Figure 2.8: *Mean standard deviation over all volunteers and slices for the three sequences (GRE, seg EPI, and ss EPI) using a TE of 17ms. Standard deviation of the temperature was calculated over 12 minutes for both field strength's (1.5 T and 3 T) within both ROI's in figure 2.2c.*

The ss EPI provided the best temperature precision on the 3 T system when using a TE near the T$_2^*$ relaxation time of brain, and did so in half the acquisition time of the GRE sequence (Table 2.3). For all three sequences tested, the precision was improved by a factor of $\approx$ 1.5 at 3 T when compared to 1.5 T (Tables 2.4 and 2.3). At both field strengths all three sequences demonstrated similar values for SNR and for standard deviation of the temperature, for TE of 17 ms and an averaging factor of 1. For all three sequences the fluctuation of the standard deviation decreased from 1.5 T to 3 T (Figure 2.7).

The SNR of the magnitude images for the three sequences GRE / seg EPI / ss EPI on the 3 T system was 63±3 / 68±3 / 65±7 and on the 1.5 T system 54±3 / 50±3 / 57±7 respectively (Table 2.4, Table 2.3). Calculated standard deviations of temperatures are consistent with Equation 2.1 for all data acquired.

Standard deviation of the temperature over space and time as well as the SNR was, on average, 20% smaller in ROI 2 (deep brain) than in ROI 1 (frontal lobe) (Figure 2.2c).

2.4 Discussion of temperature monitoring of non-moving organs

The aim of this study was to compare fast, real-time temperature imaging using the ss EPI sequence to the standard GRE and seg EPI sequences for clinical every day use in non-moving organs. The GRE sequence has been tested extensively in the past and is seen as the gold standard for real-time MR temperature mapping [112]. Temperature measurement with the seg EPI used in this study was validated previously by Cernicanu et al. [11].

To examine the temperature accuracy of the ss EPI sequence, tissue heating was performed in an ex-vivo swine model and compared to the GRE sequence. The ss EPI sequence differed only slightly from the GRE sequence, with a normalized residual of <1 °C. This small difference between the two sequences could be explained by the fact that heating of the skeletal muscle with HIFU was measured sequentially with both sequences, in different locations within the tissue. As well, slight variations in the inherent activation delay of HIFU heating may have caused this deviation. In either case, the ss EPI sequence is able to provide accurate temperature measurements which are equivalent to the standard GRE sequence.

To examine the precision of the ss EPI sequence and to compare it to the standard GRE and seg EPI sequences in non-heated volunteers' brains at both 1.5 T and 3 T. Because the ss EPI sequence is very fast, it allows for using longer TEs while still maintaining shorter acquisition times than either GRE or seg EPI sequences. This allows the use of a TE in the range of T_2^* with the ss EPI which results in a precision of 0.16 °C for a TE of 50 ms at 3 T and a precision of 0.25 °C for a TE of 70 ms at 1.5 T, without averaging (Table 2.5). In comparison, the GRE / seg EPI reach a precision of 0.37 °C / 0.39 °C at 3 T and 0.58 °C / 0.63 °C at 1.5 T with more than double the acquisition time of the ss EPI sequence. Essentially, the ss EPI provided more than twice the precision of both the GRE and seg EPI sequences at both 3 T and at 1.5 T, despite the short acquisition time of the ss EPI.

On the 3 T system, the precision of all three sequences improved by a factor of 1.5 when compared to the 1.5 T system (Figure 2.7, Table 2.4, Table 2.3). The fact that temperature mapping at 1.5 T was less precise than at 3 T can be explained by the following arguments. The relationship between the temperature precision and the macroscopic magnetic field (B_0) is proportional as determined by equation 1.60. As B_0 increases, so does the temperature precision. Furthermore, the signal amplitude of the magnitude image is doubled with the field strength, which again results in a better signal-to-noise ratio (SNR) of the magnitude image as well as a higher precision of temperature measurement. By increasing the field strength from 1.5 T to 3 T, the T_1 relaxation time, the SNR and the temperature sensitivity $\frac{\Delta \varphi}{\Delta T}$ all rise, whereas the T_2^* relaxation time decreases. Nevertheless, the parameters of the three sequences were set for 1.5 T and were not changed at 3 T in order to compare performance on both the 1.5 T and the 3 T systems. In total, all of these effects led to the demonstrated improvement in temperature precision by a factor of 1.5 in the presented data increasing the field strength from 1.5 T to 3 T. This also explains why the variance of the SD is lower for all sequences on the 3 T system when compared to a 1.5 T system (Figure 2.7). The effect of

Table 2.5: *Overview to temperature precision at 1.5T and 3T using the GRE, seg EPI, and ss EPI sequence*

		median temperature precision [°C]		acquisition time for 7 slices [s]		median SNR of magnitude image	
		1.5 T	3 T	1.5 T	3 T	1.5 T	3 T
GRE		0.55 (±0.31)	0.37 (±0.11)	7,5	7,5	54 (±3)	63 (±3)
seg EPI	TE=17 ms	0.48 (±0.39)	0.39 (±0.11)	3,3	3,3	50 (±3)	68 (±3)
ss EPI		0.54 (±0.01)	0.32 (±0.01)	1,3	1,3	57 (±7)	70 (±4)
ss EPI	TE≅T_2^*	0.28 (±0.01)	0.16 (±0.06)	3,5	2,8	94 (±3)	148 (±14)

both, the increase precision and the lower variance of the SD at 3 T, indicate a more robust temperature measurement at 3 T.

Several artifacts occur during acquisition. The most important ones are inter- and intra-scan motion, geometric distortions, N/2 ghosting and B_0-drift. Inter-scan motion artifacts can be neglected, because all results show a strong agreement between the median temperature precision over time and over space (Tables 2.4 and 2.3). Intra-scan motion artifacts have a greater influence when longer acquisition times are used, but for this experiment could be dismissed because the measurements were performed in human brain, which is essentially motionless. Even more precise temperature measurements are probably possible during clinical interventions where the head is generally fixed in place with a stereotactic frame.

Geometric distortions and N/2 ghosting along the phase encoding direction are the main artifacts affecting the EPI sequences, likely due to the sequence design [143]. In both of the EPI sequences, more k-space lines are acquired in one excitation (all of k-space for ss EPI, and 13 lines for seg EPI) than in the GRE sequence, where only one line is acquired per excitation. These artifacts are intensified in transition regions, such as between air and tissue or bone and tissue. EPI sequences have a high sensitivity to main magnetic field inhomogeneities because additional shifts accumulate within the prolonged data acquisition windows and result in artifacts that are seen as geometric distortion due to signal mis-mapping along the phase encoding direction. Nevertheless, these geometric distortions are not a safety problem for temperature monitoring during interventions, because temperature data is overlaid on the magnitude images for orientation. To achieve spatial encoding as quickly as possible, EPI acquired signals during both positive and negative x-gradient applications. This causes inconsistency in the acquired data and subsequently results in N/2 ghosts. For this study no visible artifacts by N/2 ghosting occurred.

Table 2.6: *Comparison of results from previous studies and the current study in non heated tissue.*

	Tissue	Ex vivo/in vivo	Voxel size [mm^3]	Acquisition time/slices [s]	Precision [°C]	SNR
GRE						
Meister [91]	Agarose Gel	-	1.6x1.6x8	5.1 / 1 slice	0,12	-
	Liver of a swine	Ex vivo	1.6x1.6x8	5.1 / 1 slice	0.11	-
Cernicanu [11]	Agarose Gel	-	2x2x6	6.4 / 1 slice	1.8	52
Current study	brain	In vivo	2x2x2	7.5 / 7 slices	0.58(±0.28)	37(±8)
Seg EPI						
Weidenstein [141]	liver	In vivo	3x4x5	1.2 / 3 slices	2.3 (range 1.5−5.0)	60-100
Cernicanu [11]	Agarose Gel	-	2x2x6	0.6 / 1 slice	0.5	57
	liver	In vivo	2x2x6	0.6 / 1 slice	1.3(±0.4)	33(±7)
Current study	brain	In vivo	2x2x2	3.3 / 7 slices	0.63(±0.25)	44(±13)

In comparison to the GRE and seg EPI sequences, the ss EPI sequence provides the fastest temperature measurement with the best precision. One slice with an isotropic resolution of $2\,\text{mm}^3$ length can be measured with a field of view (FOV) of $256\,\text{mm}^2$ in $0.4\,\text{s}$. Even when using a relatively long TE in the range of $T_2^\star$ ($50\,\text{ms}$ and $70\,\text{ms}$), the ss EPI sequence is still faster than the GRE sequence and in the range of the acquisition time used by the seg EPI sequences, which use short TEs ($17\,\text{ms}$). This short acquisition time reduces intra-scan motion in comparison to GRE and is one reason for the high precision of ss EPI. As well, under these conditions, the ss EPI sequence allows for 4 averages while still allowing for real-time temperature measurement. Furthermore, averaging increases the temperature precision, especially when TE is smaller than $T_2^\star$. The GRE and the seg EPI sequences do not allow for averages because of their long acquisition times. The faster imaging provided by EPI may be particularly useful for temperature measurement in moving structures, such as liver, or kidney, although the additional motion artifacts and the susceptibility artifacts would make accurate temperature monitoring more problematic.

With a temperature precision of $0.16\,°\text{C}$, measured in $378\,\text{ms}$, a detailed assessment of tissue status during an intervention is possible. If temperature measurements are done during continuous heating, a temperature dose could be calculated. This would enable a physician to make reliable predictions about condition of the tissue during an intervention in real-time, which is important both for patient safety and for evaluation of treatment efficacy.

Meister et al. [91] demonstrated that the PRF method provides a good linear correlation between data acquired with MR and those from a fluoroptic thermometry system (Luxtron, LumaSense Tech., Santa Clara, CA, USA) on a 1.5 T scanner (Sonata, Siemens Healthcare). In their study, they were able to achieve an average accuracy of $4\,°\text{C}$ to $5\,°\text{C}$ as well as an average precision of $0.1\,°\text{C}$ with the standard GRE sequence for 7 ex-vivo measurements with an acquisition time of $5.4\,\text{s}$. In contrast to the study presented in this thesis, the accuracy of the sequence was examined as a comparison between two different measurement units, Luxtron and MRI. In this thesis, a precision of $0.6\,°\text{C}$ is described over 100 recurrences in an in-vivo model using the GRE sequence, on a 1.5 T clinical scanner with an acquisition time of $7.5\,\text{s}$ for seven slices (Table 2.6). Weidensteiner et al. [141] evaluated the seg EPI for a thermo therapy at the liver of patients and volunteers without a therapy. The SNR in the liver was in the range of 60-100, the standard deviation in the range of $1.5\,°\text{C}$ to $5\,°\text{C}$ with a mean temperature SD of $2.3\,°\text{C}$. In regions with residual movement artifacts, such as areas close to the stomach or chest wall, an $SD > 5\,°\text{C}$ was observed. The seg EPI used by Weidensteiner et al. had an acquisition time of 1.2s. In some cases they used a SENSE factor of 2 and achieved TA = $0.6\,\text{s}$. The seg EPI used in the study presented in this thesis had a acquisition time of $3.3\,\text{s}$ for seven slices without parallel imaging. Further on, it may be able to further reduce motion artifacts using this sequence by applying parallel imaging techniques, and further shortening acquisition time (Table 2.6).

With the Luxtron fluoroptic thermometry system Cernicanu et al. [109] demonstrated that both the GRE and the seg EPI sequences enable safe ex-vivo temperature measurement on a 1.5 T Espree MR scanner (Siemens Healthcare). Their study was based on the same GRE and seg EPI sequences that were evaluated in this paper. The magnitude image SNR of the agar gel phantom without heating was about 60 / 55 for the seg EPI / GRE. The standard deviation of the temperature in a non-heated region of the agar gel phantom was 0.5 °C / 1.8 °C for the seg EPI / GRE. This data is on par with the data from the current study (Table 2.6).

In summary, three main messages can be derived from this study which includes volunteer data measured in the brain at 1.5 T and 3 T:

- In general, the temperature accuracy over time for the ss EPI is equivalent that of the standard temperature mapping sequences, GRE and seg EPI, for similar protocol settings. Due to its' intrinsic properties, the ss EPI sequence allows for very rapid MR temperature map acquisition in a fraction of time of the standard temperature mapping sequences.

- Investing some of the gain in speed into a longer TE (approaching T_2^*) allows a further improvement in temperature accuracy, by a factor of 1.5 at 1.5 T and by a factor of 3 for 3 T.

- Comparing the results for the ss EPI at 1.5 T and 3 T, better temperature accuracy can be achieved at 3 T (a factor 1.75) and can be acquired more rapidly as the T_2^* of the brain drops from $\propto 70\,\mathrm{ms}$ to $\propto 50\,\mathrm{ms}$ at 3 T.

In conclusion the PRF combined with the ss EPI sequence on a 3 T system offers high temperature stability, high temporal and sufficient spatial resolution when compared to GRE or seg EPI at either 1.5 T or 3 T. This makes the ss EPI sequence robust, fast and accurate enough to monitor temperatures during thermal therapy, and it may provide a greater measure of patient safety and therapeutic monitoring than current standard techniques.

Chapter 3

Temperature monitoring of the liver, as a moving and deforming organ, in patients and volunteers using the standard reference PRF method

This chapter deals with temperature monitoring using MRI of moving and deforming organs like the liver. A clinical evaluation of MR temperature monitoring of the laser induced thermal therapy has been performed in human liver using the PRF shift method. Furthermore, predictive models of cell death were examined and an exponential fit based on the Bioheat equation proposed by Pennes has been developed and evaluated.

3.1 Introduction to temperature monitoring of moving and deforming organs

MRI has the ability to increase the acceptance of local tumor thermo therapy as an alternative to conventional surgical resection. The use of multiple applicators during ablative therapies, the heat sink effect of vessels, and variable tissue perfusion make adequate monitoring of those procedures essential. Ideally real-time temperature visualization should be provided. MRI is a useful modality in conjunction with thermal ablative therapies because MRI's multi-planar capabilities allow the visualization of the placement of thermal devices in difficult locations (e.g. liver dome). As MRI also provides an excellent soft tissue contrast, it can be used to verify and characterize malignant and surrounding healthy tissue prior to, during, and after thermo therapy. Most relevant to this study, MRI is the only imaging modality that allows for non-invasive, real-time temperature mapping (TMap) during a thermal ablation procedure using a visualization of relative temperature values in the clinical setting (Section 1.3).

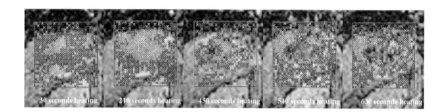

Figure 3.1: *Online temperature monitoring using standard reference PRF method. Visualization was performed as overlay of a ROI to monitor temperature and a magnitude image. Temperature maps were acquired using a GRE sequence simultaneous to LITT at human liver at several time points (blue: 35-43°C; light blue: 43-52°C; yellow: 52-60°C; red: 60-90°C).*

Nevertheless, temperature visualization in the liver is particularly challenging. It is affected by motion of surrounding lung, heart, and colon, and by the deformable nature and vascularity of the liver parenchyma itself.

The possibility of temperature measurement with a GRE sequence using the standard reference PRF method was discussed in previous studies [12, 18, 111, 112]. This study was conducted in order to examine the accuracy and precision of real-time temperature mapping during LITT in human liver using standard reference PRF method based GRE sequence (Section 1.1.5). First, the precision of the GRE sequence was evaluated in a non-heated liver of a volunteer. Second, the feasibility of simultaneous real-time temperature monitoring and LITT was examined in 18 patients with a total of 34 lesions. Third, an assessment of correlative real-time temperature mapping during LITT was performed by a comparison of the thermal dose and peak temperature to MR evidence of tumor necrosis 24 hours after LITT. The results of this patient study show the feasibility of real-time interaction dependent on temperature monitoring to control lesion size, to save healthy tissue, and to destroy all malignant tissue.

3.2 Technical setup and evaluation of volunteer and patient data

3.2.1 Technical setup

3.2.1.1 Sequence design

MR examinations were performed on a 1.5 T system (MAGNETOM Avanto, Siemens Healthcare, Erlangen, Germany) using standard body array and spine array coils. For B_0-correction, a square region covering 36 voxels was positioned in the paraspinal musculature.

For the selection of suitable GRE sequence parameters, several factors had to be taken into account. The CNR is influenced by both the echo time and the effective transverse relaxation time T_2^* (Section 1.3.2). Maximized CNR and highest sensitivity of PRF method could be acquired if TE is in the same range as $T_2^* \approx 30$ ms [2, 109, 127]. However, due to the short expiration part of the breathing cycle and the temperature changes inherent to thermal therapies, short sequence duration with high temporal resolution is desirable in order to acquire adequate temperature maps. Therefore, a compromise between short measuring time (short TE) and higher temperature resolution and accuracy of the temperature scale (long TE) is necessary. From this point of view, TE was set in combination with TR and bandwidth (BW) to TE = 12 ms, TR = 970 ms, BW = 260 Hz/voxel. The flip angle was 65° based on the Ernst angle. A field of view of 320 mm with a symmetrical matrix size of 128 and a slice thickness of 3 mm was chosen. Three parallel slices were acquired with a distance of 6 mm. All images were acquired with fat suppression, and respiratory triggering was performed using a respiratory bellow. Parallel imaging was used with a GRAPPA factor of 2, and 24 reference lines to enable an acquisition time smaller than 1 s per slice. Partial Fourier was also used to reduce the acquisition time by a factor of 75 %. Slices for temperature monitoring were acquired sequentially each in one breathing cycle.

3.2.1.2   Treatment

Thermotherapy was performed in 34 hepatic tumors (both primary and metastatic) in 18 patients (mean age 67 years). The study protocol was approved by local ethics committee and written informed consent was acquired from all patients. Patients received local anesthesia (20 ml of 1 % prilocaine subcutaneously) prior to the intervention. Prior to the procedure all patients also received a mild IV anxiolytic (Haloperidol (10 mg haloperidol; 100 mg pethinide)). All procedures were performed by an experienced interventional radiologist.

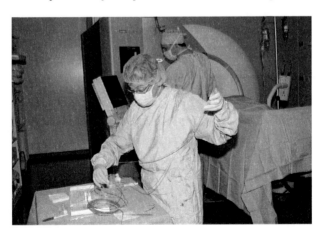

Figure 3.2: *Laser Induced Thermo Therapy (LITT) on the liver at the University Hospital in Greifswald, Germany.*

22 lesions were colorectal metastases with an average diameter of 2 cm (Table 3.1), 2 lesions were multifocal hepatocellular carcinomas, 1 a hepatocellular carcinoma, 1 lesion a metastasis of a pancreas carcinoma, 1 malignant melanoma, 1 breast carcinoma, and 2 lesions endometrial carcinomas. All 34 lesions were treated with a Neodym YAG-Laser (Neodym doped Yttrium-Aluminium-Granat-Laser, Medilas fibertom 5100, Dornier, Wessling, Germany) operating at a wavelength of 1064 nm and a maximum power of 100 W. For intervention, a micro-catheter system (Anacath, Berlin, Germany) was placed under MR guidance. This micro-catheter system consisted of a titanium needle with a tip of 1.5 mm in diameter and a surrounding transparent heat-resistant Teflon catheter with an outer diameter of 1.8 mm. After insertion, the titanium needle was replaced by a laser light guide, and thermo therapy was initiated. The MR-guided needle placement was performed interleaved using FLASH (Fast Low Angle SHot) and VIBE (Volume Interpolated Breath hold Examination) sequences with patient bed shifted out of the magnet. How often the patient bed was shifted out of the magnet depends strongly on the position of the tumor and its reachability, averaged 4 cycles were necessary. One applicator was placed in approximately 15 minutes. In a majority of 18 treatments, 2 LITT applicators were used, which were placed in adjacent parallel planes. The energy of the laser was applied uniformly over 20 minutes on average (increasing from 8 to max. 14 W, on average 20 kJ per applicator). Three parallel slices using the GRE sequence for temperature visualization were positioned parallel to the laser applicators. The goal was to place one slice in the plane of the applicators and to place the two other slices so that one was above and one below the plane of the applicators with a gap of 200 % slice thickness. In all cases the B_0-correction was placed within the nearby muscle. The LITT was controlled by the interventional radiologist based on the real-time temperature monitoring data.

In order to evaluate image quality the SNR of the magnitude was calculated in a square ROI (6 x 6 voxels) in a non-heated area of each liver. The SNR of the magnitude images was calculated using the dual acquisition method (Section 2.2):

$$\text{SNR} = \sqrt{2}\frac{S_1}{\text{SD}_{1-2}} \tag{3.1}$$

where S_1 is the mean signal intensity in a non heated ROI on the reference magnitude image in the liver, and SD_{1-2} is the standard deviation in the same ROI on a subtraction image. The subtraction image was calculated by subtracting the reference and the first temperature magnitude image. The factor $\sqrt{2}$ arises because the noise is derived from the difference image. For each measurement the SNR was averaged over the acquired 3 slices.

To estimate the influence of motion, the standard deviation of temperature was calculated over time using the same non-heated ROI used for SNR calculation.

3.2.2 Evaluation of temperature accuracy on the liver of volunteers

To evaluate the precision of the GRE sequence, repeated measurements of approximately 23 minutes (100 repeated measurements) were performed in one volunteer. For temperature mapping, three axial slices were positioned in the central part of the liver. Analysis of measurements was performed by calculating SD maps from series of consecutive temperature measurements. Therefore the standard deviation for each single voxel was calculated for all measurements. In order to evaluate image quality the SNR of the magnitude was calculated like described above using a square ROI (6 x 6 voxels) in the liver.

3.2.3 Evaluation of patient data acquired during LITT

In order to assess tumor necrosis, a follow-up MRI examination performed 24 hours after LITT was used to evaluate the TMap acquired at the end of the previous day's laser heating.

The follow-up examination included contrast-enhanced (Gadovist, Bayer-Schering, Berlin, Germany) MR-images (FLASH 2D sequence with a TR = 115 ms, TE = 5 ms, TA = 1 s, 30 slices, flip angle = 70°, no fat suppression, same voxel size as GRE sequence for the TMap). Necrosis was defined as non-enhancing (hypointense) tissue after the administration of the contrast agent.

To compare the necrotic zone in the follow-up images acquired 24 hours after LITT to the area destroyed during thermo therapy, areas of coagulation had to be defined in the TMap images. The peak temperature was used to determine and to segment the coagulated liver tissue. The peak temperature was evaluated using the temperature images from the last measurement just before turning off the laser power. In order to identify the correct peak temperature the thermal dose of Arrhenius was calculated. The first step to calculate the thermal dose was to fit the temperature data over time using the Bioheat equation as explained in the following section.

3.2.3.1 Fit of temperature data

Artifacts (e.g. inter- and intrascan motion) and noise both affect temperature mapping in human liver. Fitting the temperature images over time for each voxel allows for a more adequate temperature evaluation and reducing the artifact influence occurring only for several time points (Section 1.4.1). Fitting of phase data takes into account the information of all time points measured before and therefore minimizes the errors in any single measurement. The fit is based on the Bioheat equation Pennes proposed [16,104] to describe the temperature distribution in tissue.

Here, a small heating source is considered, that is, the infrared light penetration depth is far smaller than the specific heat diffusion length over several minutes.

An analytical solution of the Bioheat equation, corresponding to a thin cylindrical source and written for an observation point sufficiently far away from large vessels, exists for two limit cases:

- close to the fiber tip, written in rotation-invariant cylindrical coordinates (ρ, z):

$$\Delta T(\rho, z, t) = A \cdot t^{-1} \cdot exp\left(-\frac{\rho^2}{4 \cdot \kappa \cdot t}\right) \tag{3.2}$$

$$A = const \tag{3.3}$$

where κ is the isotropic thermal diffusion coefficient, ρ is the in-plane distance from the heating source, t the time, and A is a constant scaled by the diffusion coefficient.

- far away from the tip (asymptotic solution), written in spherical coordinates (isotropic solution):

$$\Delta T(r, t) = B \cdot t^{-3/2} \cdot exp\left(-\frac{r^2}{4 \cdot \kappa \cdot t}\right) \tag{3.4}$$

$$B = const \tag{3.5}$$

where r is the radial distance from the heat source origin and B is a constant scaled by the diffusion coefficient.

In an arbitrary voxel, an approximated temporal solution can be written as an intermediate situation to equations 3.2 and 3.4, and further to use it as an adjustment model in a cost-function (F) to be minimized (l_1-metrics) after each new measurement. The most recent sampling point is ranked here by the index i:

$$F_i = F(p_{1i}, p_{2i}, p_{3i}) = \sum_{m=1}^{m=i} \left| p_{1i} \cdot m^{p_{2i}} \cdot exp\left(-\frac{p_{3i}}{t_m}\right) - \Delta T_m \right| \tag{3.6}$$

where t_m is the series of time sampling at acquisition, $\vec{p}_i$ are the fit parameters after i measurements and the experimentally obtained values of temperature elevation in that voxel are denoted as ΔT_m. The fit parameters minimizing the cost function were used to calculate the model-based regularized temperature values at each time point. To evaluate the correction algorithm the chi-square (ξ^2) test was applied for each voxel in the ablation zone in each slice and patient:

$$\xi^2 = \sum_i \frac{(T_m(t_i) - T_{cor}(t_i))^2}{T(t_i)} \tag{3.7}$$

In addition to the fitting of the temperature data $T(t)$, the mean SD of temperature data and fit was calculated.

3.2.3.2 Thermal dose models

To estimate the tissue state (reversible or irreversible damage, percent denaturation), two models were assessed using the available thermal data of the fit. The first model is based on the **Arrhenius damage integral** [117] (Section 1.4.2). The Arrhenius parameters were set to $A = 9.4 \cdot 10^{104}\, s^{-1}$ and $E = 670\,\mathrm{kJ/mol}$ according to the postdoctoral lecture qualification of Ritz [116].

The second model was a **critical temperature** approach (Section 1.4.2). The peak temperature was estimated for each voxel using the thermal data of the fit. Based on prior studies the critical temperature was set to $T = 52\,^\circ\mathrm{C}$. At this temperature tissue coagulation had occurred [14, 105, 125] and any temperature above $T = 52\,^\circ\mathrm{C}$ is considered to have caused irreversible cell damage. Both models, peak temperature method and thermal dose method (Arrhenius), were compared retrospectively.

3.2.3.3 Segmentation and registration

Temperature zone $T \pm 52\,^\circ\mathrm{C}$ from the TMap was compared to the necrotic zone 24 hours after LITT (follow-up image) to examine the accuracy of real-time estimation of necrosis on the basis of temperature monitoring. First, a non-rigid registration of 3 magnitude images from the temperature mapping sequence onto post-contrast T_1 (longitudinal relaxation time) weighted follow-up images was performed. The contrast enhanced 3D dataset from the FLASH sequence was loaded in a 3D viewing tool (Siemens). The same imaging plane was located using the slice orientation information from the temperature-sensitive sequences and nearby anatomic landmarks like vessels and bones. For each slice, the TMap data and follow-up data were compared using MATLAB. The GRE temperature-sensitive images were registered on the contrast-enhanced images from the follow-up examination using the software registration tool of the MATLAB image post processing toolbox and at least 6 distinct corresponding control points. Images were transformed using a polynomial fit, cut and resized automatically. Validation of the non-rigid registration was performed by manually segmenting the whole liver for each slice of TMap and follow-up image. Corresponding segmented areas of the whole liver for each slice position were overlaid and the intersection region was calculated. After registration was completed, segmentation of the treated area in the temperature images and the corresponding post-contrast images was performed. The segmentation of the necrosis in the real-time follow-up post-contrast T_1-weighted image was performed semi-automatically by a region-growing algorithm [65].

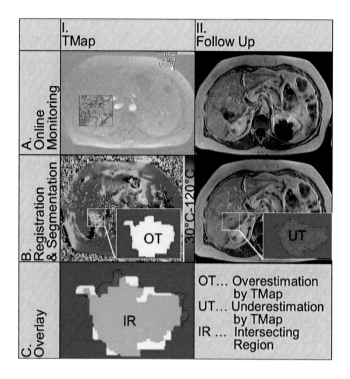

Figure 3.3: *Example of online temperature monitoring in the liver is shown in field I.A. Temperature information in field I.B. include temporal progress using the exponential fit (Method: Accuracy of real-time temperature mapping during LITT). Further on the area with temperatures $T \pm 52\,°C$ is segmented. Field II.A. shows contrast enhanced follow up image acquired 24 hours after intervention. In field II.B. a region growing algorithm was used to segment the necrotic area. The overlay of both segmented areas in field I.B. (yellow) and field II.B. (red) includes areas of overestimation by TMap (yellow, OT), areas of underestimation by TMap (red, UT) and the intersecting region (green, IR).*

The temperature region with $T \pm 52\,°C$ in the TMap image was segmented on the basis of the calculated peak temperature using a contour plot. The contour of $T \pm 52\,°C$ and segmented necrosis of the follow-up image were overlaid slice by slice for further analysis. Figure 3.3 demonstrates the evaluation procedure of the overlaid segmented structures (temperature region $T \pm 52\,°C$ and necrosis in follow-up image). The overlay is divided into 3 different areas for evaluation: OT (overestimated areas of the necrosis by the TMap), IR (intersecting region) and UT (underestimated areas of the necrosis by the TMap). To assess the under- and overestimated region by TMap in relation to the necrotic zone in the follow-up image, the ratios of the UT and the OT to the entire necrosis (UT + IR) were calculated retrospectively.

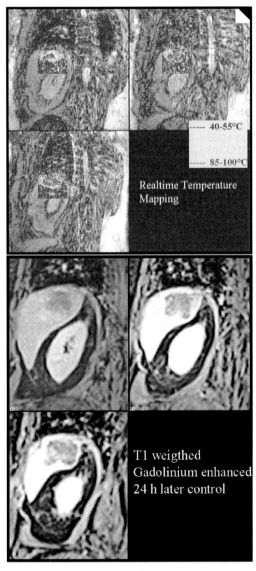

Figure 3.4: *Three separate slices acquired during LITT and 24 hours later as contrast enhanced follow up image have to be registered and segmented. The left image shows the overlay of temperature within the ROI and magnitude during LITT. The right image shows contrast enhanced follow up image 24 hours after LITT.*

3.3 Results of temperature monitoring of moving and deforming organs

3.3.1 Evaluation of the temperature accuracy on the liver of volunteers

SD maps illustrate the standard deviations in non-heated areas values for temperature in a slice over 23 minutes (Figure 3.5) and allows for an estimation of the temperature precision. A typical temperature distribution for one voxel in the non-heated liver of the volunteer is demonstrated in Figure 3.5(a). The voxel-by-voxel SD values reached 1 °C to 2 °C for all three slices over 23 minutes. Figure 3.5(b) shows a typical SD-Map for one slice. The mean SNR of the 3 slices in the liver was 18.

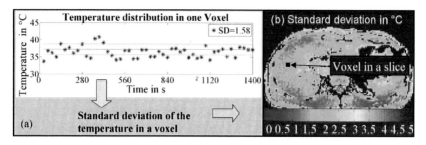

Figure 3.5: *Illustration of an SD Map: (a) Temperature distribution of one voxel with the calculated standard deviation (SD). (b) Illustration of the standard deviation of all voxels in one slice. Blue = lowest SD, Red = highest SD*

3.3.2 Evaluation of patient data acquired simultaneously with LITT

For all 34 lesions (in 18 patients), real-time temperature monitoring using the GRE with respiratory triggering was possible. No acute complications or adverse effects against Gadovist were observed.

Despite the fact that the TMap quality and the SNR differed from patient to patient, all 34 lesions were fully coagulated by LITT without MR evidence of residual malignant tissue immediately and 24 hours post treatment. Table 3.1 demonstrates the SNR of the magnitude image for each lesion acquired in a non-heated region of the liver averaged over three slices. The SNR of the magnitude images varies from 5.1 to 14.1 (mean: 9.9 ± 2.1). The standard deviation of temperature over time in the non-heated ROI varies from 1 °C to 8 °C (median: 4.9 ± 4.5 °C). For three interventions these standard deviation of temperature was above

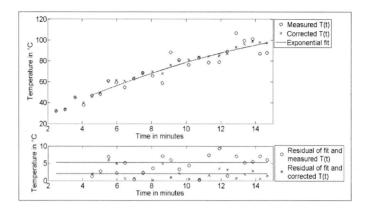

Figure 3.6: *Relative temperature data (measured with PRF method) over time, exponential fit $T(t)$, and residuals between measurement data and fit. Temperature distributions are shown for a distance of 8 mm from the laser fiber during an intervention. The exponential fit is based on Equation 3.*

$9\,^{\circ}$C. These results demonstrate that although motion is still a problem for this method of temperature mapping, the calculated temperature values were still realistic for 31 out of 34 interventions (90%).

The temperature distributions over time T(t) for the 34 LITT's were fitted for each voxel using Equation 4.47. A typical distribution of T(t) for various distances from the laser fiber can be seen in Figure 3.6. The ξ^2-test at significance level of $\alpha = 0.05$ was valid for $90\,\%$ of all voxels in the ablation zone in each slice and patient. The standard deviation for the corrected temperature data decreased by a factor of 1.9 in comparison to the standard deviation for the measured temperature data. The fits of measured temperature data over time were used to calculate the peak temperature and to calculate thermal dose on the basis of the Arrhenius damage integral. Both models of tissue damage estimation were compared by an overlay using the contours $T = 52\,^{\circ}$C and $\Omega = 1$. The peak temperature zone ($T \approx 52\,^{\circ}$C) and the contour which represents $63\,\%$ protein coagulation as defined by the Arrhenius damage integral, demonstrated 97% agreement. Peak temperatures were used to compare the region $T \geq 52\,^{\circ}$C in the last image of heating with the necrotic zone assessed 24 hours later.

The non-rigid slice by slice registration between TMap and follow-up image was verified by a segmentation of the whole liver in both images. The intersection region of both whole segmented livers was approximately $95\,\%$. The comparison of necrosis assessed 24 hours after LITT and contour of peak temperature ($T = 52\,^{\circ}$C) in the TMap are shown in Table 3.1. The average intersecting region of TMap and necrosis was $87.2 \pm 5\,\%$ (range: $73\,\%$ to $94\,\%$), the averaged underestimation of necrosis by TMap was $12.8\pm5.4\,\%$, and overestimation of necrosis by TMap was $13.2 \pm 3.6\,\%$. The TMap of three lesions were not adequate for segmentation and registration.

Table 3.1: *Summary of patients, lesions, necrosis, number of applicators, applied energy, mean SNR of phase and magnitude image, UR, OR, IR for three slices in each lesion.*

Patient	Lesion	Type of lesion	Lesion Location	lesion volume [mm]	necrotic volume [mm]	number of applicators	applied energy [kJ]	SD of T non-heated [°C]	SD of T heated [°C]	SNRmagn space	UR [%]	OR [%]	IR [%]
A	1	CM	7	5x5x5	31x18x20	1	18.4	4.9	8	12.3	7	10	93
B	2	CM	5	33x20x25	45x31x28	2	37.8	4.8	12	10.2	12	17	88
C	3	HCC	7	12x12x20	48x22x46	2	44.1	7.2	13	10.6	13	17	87
	4	HCC	5	12x12x12	48x22x20	2	24.7	8.6	19	7.7	28	23	72
D	5	CM	8	42x30x25	55x40x38	2	43.4	1.8	14	14.1	7	10	93
	6	CM	6	28x25x20	61x34x46	3	65.2	6.4	4	7.9	24	17	76
	7	CM	6	10x10x10	42x25x16	2	37.7	0.9	2	8.6	15	9	85
	8	CM	8	19x18x18	48x34x30	2	27.9	3.9	13	11.3	10	15	90
E	9	CM	4	20x15x19	62x31x34	2	41.9	10.2	19	6.5	-	-	-
	10	CM	8	4x4x42	47x45x45	4	62	10	19	6.5	-	-	-
	11	CM	8	40x38x38	47x40x40	4	64.2	6.8	14	7.2	18	13	81
	12	CM	5	25x21x18	35x34x32	2	31.9	5.9	15	7.3	17	17	82
F	13	CM	5	18x15x13	48x38x25	2	24.1	3.8	9	8.5	16	16	84
	14	CM	6	24x22x20	40x39x30	2	32	6.7	14	9.5	15	13	85
G	15	EC	6	28x25x25	60x36x30	3	48.1	4.3	5	8.8	17	17	83
	16	EC	6	25x20x20	35x35x35	2	32.4	7.1	14	12.3	7	17	93
H	17	CM	6	13x10x10	44x36x40	2	43.5	4.3	14	8.8	14	17	86
	18	CM	6	16x15x15	48x27x15	2	34.8	4	14	8.8	10	17	93
	19	CM	2	21x20x20	41x21x21	2	32	2.9	5	12.3	7	10	92
I	20	CM	2	17x15x15	38x20x20	1	16	7.2	11	7.8	14	12	85
	21	CM	6	14x10x10	30x16x15	1	12.2	5.3	6	11.7	10	10	90
J	22	CM	6	22x14x14	44x32x30	2	32	3.3	6	9.4	7	10	93
	23	CM	2	18x18x18	28x27x27	2	32	2.7	8	10.4	10	12	89
K	24	CM	4	6x6x6	30x15x15	1	16	6.7	14	9.5	17	17	83
	25	CM	2	20x15x15	35x30x22	2	32	5.9	12	10.9	13	10	87
L	26	CM	4	19x19x19	20x34x46	2	31	5.9	11	9.1	17	17	83
	27	PC	3	15x15x15	36x26x25	1	18	3.8	8	13.2	7	11	93
M	28	CM	2	17x15x15	47x30x35	2	29.5	9.8	21	5.1	-	-	-
N	29	CM	2	25x21x21	53x34x30	3	44.3	6.4	13	9.3	17	17	83
	30	CM	4	12x10x10	40x29x30	2	32.1	2.8	7	12.3	7	8	93
O	31	MM	6.7.8	38x35x30	45x39x35	3	48	7.3	12	9.8	10	16	90
P	32	MM	6	32x30x31	45x35x35	3	36.3	3.2	8	11.1	8	9	92
Q	33	GC	8	56x50x50	53x45x55	4	64	2.9	8	10.7	9	9	93
R	34	HC	6.7	34x30x27	52x38x40	4	64.3	3.1	5	11.8	7	10	93
median						4		6	11	10	13	13	87
								±4	±5	±2	±5	±4	±5

The percentage under- and overestimation was calculated using the the peak-temperature of T = 52 °C in comparison to the necrotic zone in correlative contrast-enhanced T1-weighted MR images 24 hours after intervention. Lesion location refers to the liver segment.(CM = colorectal metastases, HCC = multifocal hepatocellular carcinoma, HC = hepatocellular carcinoma, EC=Endometrial carcinoma, MC = Mamma carcinoma, MM = malignant melanoma, PC = Pancreas carcinoma, GC = gastric carcinoma, Lesion location = liver segment, SNRph time = signal to noise ratio of phase-difference images over time, SNRph space = signal to noise ratio of phase-difference image over space, SNR magn = signal to noise ratio of magnitude image over space, U = region underestimated by TMap, OR = region overestimated by TMap, IR = intersecting region of necrosis and TMap)

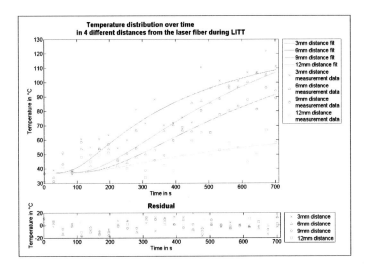

Figure 3.7: *Temperature distribution over time for three distances (3 mm, 6 mm, 9 mm) from the laser fiber. All three temperature distributions are fitted using the exponential fit based on the Bioheat equation proposed by Pennes.*

3.4 Discussion of temperature monitoring of moving and deforming organs using standard reference PRF method

Until now, no real-time temperature monitoring during LITT at human liver for a clinical patient study has been published. The aim of this study was to evaluate the clinical utility of GRE sequence for real-time temperature mapping and monitoring for LITT in the liver, a particularly challenging environment for temperature mapping, due to the liver's mobility and deformation. In the presented study the precision of GRE temperature mapping in a non-heated volunteer liver is examined, as well as the feasibility and accuracy of real-time GRE temperature mapping in a clinical study performed with patients undergoing LITT of liver tumor(s).

The precision of GRE TMap was calculated with volunteer measurements in the liver to be ≈ 2 °C. The standard reference PRF method of temperature mapping calculates a relative temperature as a difference from the reference and the present phase image, identical voxels from the reference and heating tissue must be subtracted for an accurate temperature assessment. This is particularly challenging in the liver because the liver is highly deformable, changes position throughout the respiratory cycle, and has image quality which is influenced by motion of surrounding organs, such as heart and bowel. The fact that a precision of approximately 2 °C was obtained using image triggering by respiratory bellows over 23 minutes

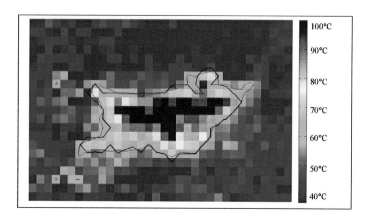

Figure 3.8: *Overlay of the isotherm of $T \pm 52\,°C$ (red line) and the contour of the Arrhenius damage coefficient $\Omega = 1$ (black line) on a temperature map corrected by Pennes' Bioheat equation. The temperature map shows the laser-induced ablation zone after 20 minutes of heating in human liver.*

makes this technique clinically applicable.

In volunteer measurements the mean magnitude SNR of 18 is due to the relatively short TR and long TE selected in order to maximize the quality of temperature mapping. The SNR would increase by increasing TR and keeping TE constant, but this would cause more pronounced motion artifacts. The low SNR can be improved by faster scan techniques and improved sequence design.

TMap data from the healthy volunteer (SD $\approx 2\,°C$) demonstrate better image quality compared to the data acquired from patients who underwent LITT (SD $\approx 5\,°C$). This can be explained by the fact that the patients, unlike the volunteers, were undergoing a percutaneous intervention, and, despite local anesthesia, had some pain and thus some irregular breathing during the procedure. This irregular breathing can lead to inter- and intra-scan motion artifacts, which typically result in errors of the calculated temperatures. The irregular breathing and resultant irregular liver displacement and deformation mean that exactly the same slice of the liver with the same deformation could not always be measured in both the reference image and the phase image that contains the information regarding the temperature elevation. Depending on the degree of displacement this causes either under- or overestimation of temperature.

In order to minimize the influence of artifacts, a fit of temperature over time $T(t)$ was applied to the data points measured with the GRE sequence. Deviations can be observed for voxels including data outliers due to motion and deformation artifacts or due to the heat sink effect nearby vessels. Nearby vessels and temperature abnormalities related to breathing artifacts could further deform the shape of the fit $T(t)$ and influence the calculated temperature maximum or thermal dose.

The fit $T(t)$ (Equation 4.47) for each voxel is the basis of peak temperature and thermal dose calculation. The averaged agreement of the peak temperature zone with $T \approx 52\,°C$ at the end of the LITT and the zone with a thermal dose leading to denaturation of more than $63\,\%$ of proteins (Equation 1.81) was greater than $97\,\%$.

Arrhenius' theory has been evaluated in various studies since 1984 [14, 105, 125]. The agreement between the Arrhenius model and the peak temperature means that an interventional radiologist can estimate the necrosis of malignant tissue in real time, based on the TMap during the intervention. Peters et al. [105] analyzed three models of thermal damage: the peak temperature, the Arrhenius damage integral, and the time temperature product during the LITT at two in vivo canine prostates. Their study demonstrated consistent histological correlation for the peak temperature of $T = 51\,°C$. The reported results support the validity of the results presented in this thesis.

The mean percent agreement of necrotic zone 24 hours after LITT and TMap zone with peak temperature $T \geq 52\,°C$ was $87\,\%$. The error of $13\,\%$ may have several causes, which can be devided into two different categories, the post processing step and temperature acquisition itself. Errors occurring as a result of the post processing steps are misregistration, missegmentation, and fit $T(t)$ (Equation 4.47). Those resulting from temperature acquisition include: a mismatch between reference and TMap images (e.g. irregular breathing, non-rigid deformation of the liver), non-corrected intrascan motion artifacts (e.g. blood flow, respiratory motion, cardiac motion), and difficulties in the positioning of the ROI for B_0-correction. The error due to the non-rigid registration was quantified by segmentation of the whole liver in both registered images (TMap and 24 hours post-contrast). The over- and underestimated regions are on average $5\,\%$. This deviation was likely due to the registration technique, which was performed on 2D images rather than on 3D volumes. A further source of deviation is the fit T(t) (Equation 4.47) of the temperature data, which served as the basis of the comparison to the necrosis 24 hours after LITT.

Nevertheless, the greatest source of error was probably liver motion and the non-rigid deformation of the liver. These factors made slice and voxel matching challenging. As well, there are significant regional variations in motion within the liver. For example, the liver dome was most affected by respiratory motion. Measurements near the edge of the liver were affected by motion of organs such as the heart and the bowel. Also, due to the sensory innervation of the liver capsule, thermal therapy adjacent to the edge of the liver causes significantly more pain than does therapy in other parts of the liver. This would likely induce further respiratory and motion irregularities. Finally, the patients' emotional state immediately prior to and during the intervention varied, further altering the respiratory rate and regularity.

The SNR of the magnitude image is also influenced by the artifacts mentioned above and therefore is an indicator for the quality of TMap as well as the temperature standard deviation over time in a non-heated region. As shown by 3.1, lesions (4, 6) which were imaged with a low SNR (7.7 and 7.9) and high SD ($8.2\,°C$ and $6.4\,°C$) demonstrated the low agreement between TMap and necrosis ($72.3\,\%$ and $75.5\,\%$), while lesions (5, 19, 27, 30, 33) with higher SNR (14.1, 12.3, 13.2, 12.3, 10.7) and low SD ($1.8\,°C$, $2.9\,°C$, $3.8\,°C$, $2.8\,°C$, $2.9\,°C$) demonstrated the best agreement (93 %) between TMap and necrosis. These results demonstrate the need for further pulse sequence improvements, primarily in speed and SNR. However, the results presented in this thesis show that the used temperature monitoring delivered the induced necrosis, seen 24 hours after intervention.

Three lesions of two patients could not be evaluated due to the poor TMap quality. The reasons for the degraded data are likely a combination of all the possible error sources. Patients were agitated, had very irregular respiration, the lesion were in the liver dome, and the B_0-correction positioning was very difficult. Overall the SNR and the TMap quality were strongly dependent on the patient and his/her ability to cooperate. Inter- and intrascan motion artifacts resulted in a poor temporal and spatial SNR of the phase images. Despite low SNR in native GRE data, using the fit with the analytical solution of the Bioheat equation effective accuracy of MRI was largely improved.

To reduce the problems induced by liver motion several methods have been proposed, e.g. multi-baseline and reference-less or self-reference approaches. Butts developed a triggered, navigated, multi-baseline method for temperature measurement using the PRF method [135]. Nevertheless, there are significant inter-view motion and blurring (vessels) artifacts. Reference-less and self-reference methods approximate the phase in a defined ROI by a polynomial [68,113]. A border around the heated zone is used to determine the polynomial coefficients by the least-square algorithm, which is used to estimate the non-heated background phase (reference phase). It is difficult to find a continuous border both without susceptibility artifacts and without heating around many lesions primarily because the numerous large blood vessels within the liver and the surrounding tissue (lung, stomach) induce susceptibility artifacts. These approaches have to be tested further before in-vivo thermo therapy at the human liver can be performed.

Instead of all these possible sources of error and deviation, which could occur during temperature visualization using the GRE sequence and PRF method, a 13 % difference between the TMap Zone of $T \geq 52\,°C$ and the necrotic zone measured at 24 hours post-intervention is considered to be a good result for real-time, temperature measurement performed without using general anaesthesia and artificial respiration. Furthermore, the quality of the temperature monitoring data was sufficient for accurate monitoring and follow-up of the therapy by the physician, as demonstrated by the fact that all 34 of the lesions addressed in this study were fully treated. Although this thesis investigated thermal therapy using LITT, which produces no artifacts on MR imaging, it would also be feasible to use the same methods described in this paper for MR-guided RFA [110]. For RFA, the induced needle artifacts would interfere with temperature mapping in zones closest to the needle, but in most cases these artifacts would not extend to the outer zones of the TMap, where the edge of the necrotic zone needs to be assessed. For HIFU, these techniques could also be used to measure the extent of the necrotic zone if the spatial in-plane-resolution is increased, even though HIFU induces rapid temperature elevations and requires real-time motion adaptation of the HIFU beam to accurately treat mobile organs [27]. The temporal resolution, which is necessary to monitor HIFU could be performed for organs not influenced by respiration.

In conclusion this study demonstrates that using a GRE sequence for MR temperature mapping based on the PRF method, an adequate estimation of tumor necrosis can be achieved during LITT treatment even for difficult conditions like the liver on the basis of a clinical patient study. By allowing the radiologist to monitor thermal therapy in real-time MR-based temperature mapping may potentially allow the physician to alter treatment in real-time according to the developing treatment conditions, namely the development of necrosis. Deviations due to motion were reduced by respiratory triggering, and produced data which was clinically acceptable. Further steps to minimize artifacts due to inter- and intra-scan respiratory motion are presented in Chapter 4.

Chapter 4

Reference-less temperature monitoring using the PRF method

4.1 Challenges in temperature monitoring of moving and deforming organs using the PRF method

MR temperature mapping based on the water PRF shift effect uses the complex signal acquired from the GRE sequences [58] (see 1.1.5) and has become an accepted technique to monitor thermo therapies [17, 18, 66, 87, 109, 111, 112] (see 1.3). The method has two fundamental advantages: tissue-independence and a linear calibration from 0 °C to 100 °C [53]. The relative temperature can be calculated from the MR phase difference between one phase image (acquired before the treatment) and one phase image that contains the information regarding the temperature elevation. The standard implementation of MR Thermometry (MRT) requires a double scan through the volume of interest, with the same organ and slice position and with the same organ deformation. These two time-points, which are necessary for the phase subtraction, may have a temporal separation of the order of even 10 to 30 minutes. Despite the use of a breathing belt [78, 130, 135], pencil-beam navigators [115] or other encoding and triggering methods for the respiratory motion, the voxels in the acquired slice are always, to a certain degree, shifted between the acquisition of the reference phase and the later acquisition during the thermal treatment. Per-operatory swelling or shrinking of the heated tissue or adjacent tissues can also be expected. The occurrence of these effects, inter-scan motion and swelling/shrinking of tissue, between the acquisition of the reference phase and the acquisition of the phase image that contains the information regarding the temperature elevation, lead to MR thermometry errors (over- or under-estimation of the calculated temperature).

Several approaches have been published that aim to reduce or eliminate these errors. Vigen et al. [135] and Sennevall et al. [27,29] discussed a multi-baseline (or equivalently, multi-reference) method. The multi-baseline approach acquires multiple reference phase images for different positions of the moving organ (e.g. preparatory mapping of the motion as a data set atlas),

and generates the per-operatory temperature maps by choosing the images with the best correlation to be used for the phase subtraction. Although multi-baseline MRT was demonstrated
to reduce inter-scan motion artifacts, it cannot completely eliminate them in living tissue as
the reference-less method does. It is practically impossible to scan exactly through the same
anatomical plane during two time-point acquisitions on moving organs, as for example the
liver, which is moved and deformed by respiration, heart beat and peristaltic of the bowel and
undergoes a slow drift [140]. Multi-baseline acquisition cannot improve accidental motion artifacts (e.g. involuntary muscular contraction induced by per-operatory pain, postural change
etc), as the required information has not been mapped before the event occurs.

Rieke et al. [114] were the first to introduce a reference-less PRF MRT method using spatial
information (instead of temporal information) to isolate the temperature-induced PRF phase
shift. The principle of reference-less MRT is to split a single time-point phase map into: 1.
the spatial reference (also called the background phase), and 2. the temperature-induced PRF
phase contrast. Rieke's initial implementation [114] used a polynomial interpolation (with l_2
metrics; typically up to the order of 6) to estimate the background phase from a rectangular,
closed, thick border toward the inner ROI. The calculated background phase (instead of the
temporal reference) was then subtracted from the actual phase map per-operatory, i.e. single
time point acquisition. Grissom at al. [47] described a polynomial variant of the reference-less
MRT implementing l_1-metrics to adjust the fit coefficients.

Reference-less MRT is intrinsically insensitive to tissue swelling/shrinking and inter-scan motion. Furthermore, reference-less MRT permits greater flexibility during the ablation procedure compared to temporally referenced MRT. The sequence parameters, as well as the slice
position and orientation, can be varied in real time during the intervention, and any susceptibility artifact-free needle may be repositioned or additional applicators may be placed. The
reference-less MRT approach described by Rieke et al. opened a new perspective for PRF-
based MRT, nevertheless, their mathematical algorithm may lead to numerical singularities
such as an ill-conditioned system of equations.

The first section of this chapter recalls the physical foundations and pre-clinical validations of
the near-harmonic reference-less method originating from the work of Dr. R. Salomir, Dr. M.
Viallon, Dr. J. Roland and Dr. P. Gross complemented by the the author of this thesis with
a real-time numerical implementation and a clinical study validation with LITT ablation in
liver.

Taking a 3D domain with no sources of magnetic susceptibility (or with linearly varying susceptibility), the unwrapped phase of the GRE 3D data is demonstrated to be a harmonic function,
which results in a null scalar field when the Laplacian operator is applied. Susceptibility inhomogeneities, localized heating, mixed signal from different chemical shift species with spatially
varying ratios, or other artifacts, are all non-harmonic functions which contribute to the globally measured GRE phase [122]. Calculating the background (reference) phase thus means
calculating the harmonic part of the phase map. The non-harmonic part of the GRE phase,
in a homogeneous medium, constitutes the temperature elevation. Phase information along
a one pixel thick border is theoretically sufficient to calculate the full harmonic background

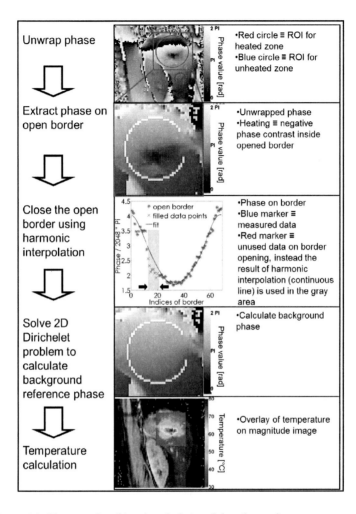

Figure 4.1: *Five steps describing the calculation of the reference-less temperature map.*

phase inside the domain (i.e. inner Dirichlet problem). This border has to be unheated and free from artifacts and susceptibility contrast. To enable slice-per-slice monitoring of temperature, the calculation of the background phase was reduced from 3D to 2D with a residual term in the phase's Laplacian of order zero. This constant Laplacian approximation is called here a near-harmonic function. To determine the residual term of the Laplacian operator, the border has to be at least two pixels thick. For practical reasons (mainly noise-robustness) a three pixels thick border is implemented for the study presented in this thesis. Moreover, openings in the border were compatible with the method by considering a circular geometry of the domain. The open border implementation permits the elimination of areas of conflict from the assigned border, such as heating, artifacts and susceptibility contrast. Firstly, the open border was re-filled to a closed border using harmonic interpolation (similar to partial Fourier techniques), which intrinsically excluded temperature-induced non-harmonic phase shift contributions over the repaired segments. Secondly, the information of the virtually closed border was used to solve the inner 2D Dirichlet problem, generating the background phase as output. Thirdly, the background phase was subtracted from the acquired phase map. The main steps involved in calculating the reference-less temperature are illustrated in Figure 4.1.

The whole reference-less MRT described in this paragraph was implemented in C++ in the so called ICE framework. This ICE framework is a vendor specific MR Image Calculation Environment (ICE), which can be used for patient treatment in a clinical environment. The reference-less MRT was tested first ex-vivo using an HIFU setup (see 1.2.3), second in-vivo using HIFU at sheep livers (see 4.3.1.2), third in-vivo at the liver of volunteer (non-heated) (see 4.3.1.3), and fourth at the liver of patients during an intervention (4.3.2).

4.2 Theoretical basics of reference-less temperature monitoring

According to the quasi-static approximation of Maxwell's equations, the magnetic field intensity $\vec{H}$ is related to the free current density $\vec{j_c}$ as:

$$\vec{\nabla} \times \vec{H} = \vec{j_c} \qquad (4.1)$$

The average free current density in biological tissue at the spatial scale of one voxel (e.g. millimeter range) is considered to be zero. Eddy currents (vortices) induced in the tissue by MRI gradients are further neglected, and the validity of this approximation is discussed in section 4.2.4. Under such conditions the magnetic intensity is an irrotational vectorial field (i.e. no eddy currents exist in the field structure) and a scalar field called "magnetic potential" Φ exists such that:

$$\vec{H} = \vec{H_0} - \vec{\nabla}\Phi \qquad (4.2)$$

where $\vec{H_0}$ denotes the static magnetic field produced by the superconductive magnet, that is highly uniform and considered as time-independent (at the time scale of one image acquisition). The following relationship exists (see for complete demonstration reference [122]):

$$\vec{\nabla}^2\Phi = H_0 \cdot \frac{\partial \chi}{\partial z} \qquad (4.3)$$

with

$$H_0 = \left| \vec{H}_0 \right| \tag{4.4}$$

$$\vec{\nabla}^2 = \left(\frac{\partial^2}{\partial x^2} + \frac{\partial^2}{\partial y^2} + \frac{\partial^2}{\partial z^2} \right) \tag{4.5}$$

Given the expression for the signal phase acquired with an RF-spoiled GRE sequence [129] (section 1.3):

$$\varphi = \mu_0 \cdot \gamma \cdot TE \cdot \left(1 - \sigma + \frac{\chi}{3} \right) \cdot H_0 \tag{4.6}$$

The bulk magnetic susceptibility of tissue is denoted as χ and the chemical shift for water protons is denoted as σ. Considering that the temperature is uniform (i.e. σ is not spatially dependent) one can relate the Laplacian of the background phase map to the magnetic susceptibility as follows:

$$\vec{\nabla}^2 \varphi = \mu_0 \cdot \gamma \cdot TE \cdot \vec{\nabla}^2 \left[\left(1 - \sigma + \frac{\chi}{3} \right) \cdot H_0 \right] \tag{4.7}$$

$$= \mu_0 \cdot \gamma \cdot TE \cdot \vec{\nabla}^2 \left[\frac{\chi}{3} \cdot H_0 - \frac{\partial \Phi}{\partial z} \right] \tag{4.8}$$

$$= \mu_0 \cdot \gamma \cdot TE \cdot \left[H_0 \cdot \frac{\vec{\nabla}^2 \chi}{3} - \frac{\partial}{\partial z} \left(\vec{\nabla}^2 \Phi \right) \right] \tag{4.9}$$

$$= \mu_0 \cdot \gamma \cdot TE \cdot H_0 \cdot \left(\frac{\vec{\nabla}^2}{3} - \frac{\partial^2}{\partial z^2} \right) \chi \tag{4.10}$$

In such regions where the magnetic susceptibility satisfies the condition:

$$\left(\frac{\vec{\nabla}^2}{3} - \frac{\partial^2}{\partial z^2} \right) \chi = \frac{1}{3} \left(\frac{\partial^2}{\partial x^2} + \frac{\partial^2}{\partial y^2} - 2 \cdot \frac{\partial^2}{\partial z^2} \right) \chi = 0 \tag{4.11}$$

e.g. regions with homogeneous or linearly-changing susceptibility, it results from equation 4.10 that the three dimensional spatially-unwrapped GRE phase map is a harmonic function:

$$\vec{\nabla}^2 \varphi = \left(\frac{\partial^2}{\partial x^2} + \frac{\partial^2}{\partial y^2} + \frac{\partial^2}{\partial z^2} \right) \varphi = 0 \tag{4.12}$$

While property 4.12 is valid in 3D, the problem has to be reduced to a 2D situation. It is far easier to deal with 2D acquisition (e.g. single slice) than with 3D acquisition when addressing real time MR thermometry, in particular in moving organs. Considering the in-plane continuous Cartesian coordinates of the slice denoted as (x_s, y_s) the second derivative along the third dimension shall be approximated by a constant value denoted as $-\epsilon$, to be determined. Determining the 2D background phase $\varphi^{bk}(x_s, y_s)$ is equivalent to solving the Dirichlet-type problem:

$$\begin{cases} \left(\frac{\partial^2}{\partial x_s^2} + \frac{\partial^2}{\partial y_s^2} \right) \varphi^{bk} = \epsilon = \text{const} & : \quad \text{inside a connex domain } \Omega \\ \varphi^{bk}(\partial \Omega) = \varphi^{exp}(\partial \Omega) & : \quad \text{on unheated closed border } \partial \Omega \end{cases} \tag{4.13}$$

where φ^{exp} stands for the measured phase map. The average residual value of the 2D phase's Laplacian (ϵ) can be easily determined for some particular configurations. R and Θ represent the Polar coordinates. For instance, considering the average value of the phase map along a circle of radius R as:

$$\langle \varphi^{\text{exp}}(R) \rangle = \frac{1}{2\pi} \int_0^{2\pi} \varphi^{\text{exp}}(R, \Theta) d\Theta \tag{4.14}$$

one can demonstrate that for two concentric circles of radii R_0 and $R_0 + \Delta R$ the following relation is valid (see section 4.2.3) :

$$\epsilon = 4 \cdot \frac{\langle \varphi^{\text{exp}}(R_0 + \Delta R) \rangle - \langle \varphi^{\text{exp}}(R_0) \rangle}{(2 \cdot R_0 + \Delta R) \cdot \Delta R} \tag{4.15}$$

The recommended value for ΔR are two voxels, a compromise between the border thickness (equal to $\Delta R + 1$ voxel) and the robustness of the contrast-to-noise ratio in the calculation of ϵ. The value of R and the center of the circle are both arbitrary so far as both circles are comprised inside a homogeneous susceptibility region. In this thesis, the circle of radius R shall be called "the inner border" and the circle of radius $R + 2$voxel the "outer border".

4.2.1   Iterative numerical solution for Dirichlet problem

The first step for the numerical implementation of the method is to spatially unwrap the experimentally measured phase map for 0 to 2π jumps. For this purpose a domain-growing algorithm has been used, which starts with the voxel with the highest magnitude signal inside the phase ROI to be unwrapped.

To solve the Dirichlet problem of equation 4.13 a fast and robust numerical method is used. In order to avoid mathematical operations which could lead to instability and/or singularity, for instance the calculation of inverse matrices, it is preferable to use an iterative approach. Based on the 5-voxel discretization of equation 4.13 using 2D indexes (i, j):

$$\varphi^{\text{bk}}_{i-1,j} + \varphi^{\text{bk}}_{i+1,j} + \varphi^{\text{bk}}_{i,j-1} + \varphi^{\text{bk}}_{i,j+1} - 4 \cdot \varphi^{\text{bk}}_{i,j} = \epsilon \tag{4.16}$$

one can propagate the solution from iteration (n) to iteration (n+1) as follows:

$$\varphi^{\text{bk}}_{i,j}(n+1) = \begin{cases} \frac{\varphi^{\text{bk}}_{i-1,j}(n) + \varphi^{\text{bk}}_{i+1,j}(n) + \varphi^{\text{bk}}_{i,j-1}(n) + \varphi^{\text{bk}}_{i,j+1}(n)}{4} - \frac{\epsilon}{4} & : & \text{inside } \Omega \\ \varphi^{\text{exp}}_{i,j} & : & \text{on the border mask } \partial\Omega \\ 0 & : & \text{outside } \Omega \end{cases} \tag{4.17}$$

The border condition is used at each iterative step so that the values of the reconstructed background phase for the voxels, which constitute the border, are reset to the experimentally measured values. The iteration is repeated until no further significant change occurs with further iterations. The output of the iterative reconstruction algorithm is used as the background

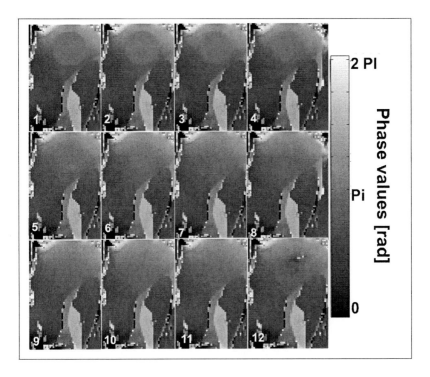

Figure 4.2: *This image visualizes the iterative reconstruction method of the background phase inside the domain of interest. The shown background calculation uses a circular border placed on in vivo liver data acquired during the LITT of a patient. The actual background phase during the reconstruction process is shown before the first iteration in frame 1) (i.e. border information only), and after 1, 5, 10, 20, 40, 100, 200, 500, 700 and 900 iterations in frames 2), 3), 4), 5), 6), 7), 8), 9), 10) and 11) respectively. Whereas, the border data has experimental noise, the reconstructed background is a smooth function. Frame 12) shows the original, unwrapped phase.*

phase, to be subtracted from the experimentally measured phase inside the domain:

$$\varphi^{\mathrm{bk}} = \lim_{n \to \infty} \varphi^{\mathrm{bk}}(n) \tag{4.18}$$

The result of that phase subtraction is proportional to the temperature variation:

$$\Delta T = \frac{\varphi^{\mathrm{exp}} - \varphi^{\mathrm{bk}}}{\alpha \cdot \gamma \cdot B_0 \cdot TE} \tag{4.19}$$

where $\alpha \cong -0.01 \frac{ppm}{^\circ C}$.

4.2.2 Extension to open border

In some situations it is not possible to provide an unheated closed border for the calculation of the harmonic background phase. Some typical conflicts occur in MR guided HIFU applications when imaging a plane along the main direction of propagation of the acoustic beam, or when monitoring volumetric sonication [101, 123]. An extension of the method is described here for cases where some segments along the border cannot be used as initial information for the Dirichlet problem. A preliminary step of "repairing" the border is necessary, that is, recovering the missing information. It is a fundamental property that a 2D near-harmonic function (i.e. constant Laplacian, see equation 4.13) considered along a circle can be written as an exact Fourier series of 1D trigonometric functions:

$$\varphi^{\text{bk}}(r = r_0, \Theta) = \alpha_0 + \sum_{k=1}^{\infty} [\alpha_k \cdot exp(i \cdot k \cdot \Theta) + \alpha_k^* \cdot exp(-i \cdot k \cdot \Theta)] \qquad (4.20)$$

The star superscript denotes the complex conjugate ($\varphi^{\text{bk}}(r = r_0, \Theta)$ is a real function). For practical implementation, the maximum rank of the Fourier coefficients is limited to a value denoted as N:

$$\varphi^{\text{bk}}(r = r_0, \Theta) \cong \alpha_0 + \sum_{k=1}^{N} [\alpha_k \cdot exp(i \cdot k \cdot \Theta) + \alpha_k^* \cdot exp(-i \cdot k \cdot \Theta)] \qquad (4.21)$$

The open border can be re-filled to constitute a closed border using an iterative reconstruction based on successive direct and inverse 1D Fourier transformations. This is equivalent to finding the best set of α_k coefficients using only the phase information from segments of the border included in the calculation. The value of N should be defined in order to compromise between taking into account the local high frequencies whilst not reaching unstable behavior on the open portions of the border. The implementation developed in this thesis uses a semi-empirical criterion for self limiting the maximum rank of the Fourier coefficients. The reconstruction of the border starts at $N = 3$. N is then sequentially incremented by one unit as long as following condition is respected:

$$|\alpha_{N-2}| < |\alpha_N| \qquad (4.22)$$

The iterative algorithm used to determine the coefficients α_k avoids possible singularities and works as follows (see Figure 4.3 for illustration):

1. The missing values from the border are linearly interpolated from the closest known edges generating the first approximation vector $\varphi^{\text{bk}(0)}(r = r_0, \Theta)$, see Figure 4.3.a

2. The coefficients α_k up to order N are determined from a standard Fourier transformation of the current series of phase values on the border; note that coefficients α_k of higher order than N are set to zero, see Figure 4.3.b

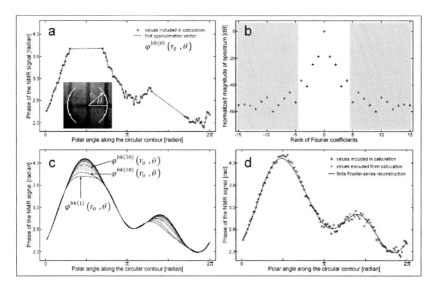

Figure 4.3: *Unheated-heated test case for iterative reconstruction of the phase along the open segments of the border (ex vivo sample). a) Data was considered to be available along a total of 60% of the voxelized border ($r_0 = 20$ voxel radius circular domain), with 40% "missing data". The encrusted image illustrates the geometrical frame. The phase values considered for test purposes as "unavailable" were initially set as the linear interpolation from the edges, i.e. the approximation of order 0; b) Magnitude of the Fourier spectrum of the initial approximation. The coefficients falling inside the shadowed areas are set to zero before inverse Fourier transformation of the truncated spectrum; c) Evolution of the iterated approximations, the reconstructed values of phase are each time reset to the experimental values along the existing segments of the border, before direct Fourier transformation and spectrum truncation. For illustration purposes, iterations of order 1, 10 and 50 respectively are highlighted; d) The asymptotic solution compared to the ground-truth phase values which were actually known in this test case. The quality of the phase reconstruction on open segments expressed in equivalent °C at the given echo time at 3T was 0.15°C accuracy and 0.71°C precision.*

3. The spectrum truncated to order N is inverse Fourier transformed generating the next approximation vector $\varphi^{\text{bk}(n)}(r = r_0, \Theta)$, see Figure 4.3.c.

4. In the current approximation vector $\varphi^{\text{bk}(n)}(r = r_0, \Theta)$, the voxels from the useful border segments are reset to the experimentally measured phase values

5. Steps 2 to 4 are repeated until convergence is reached (e.g. end-point), see Figure 4.3.c and 4.3.d.

This algorithm works for an arbitrary number of openings on the border, provided that none of the openings exceed a quarter of the circle and that the accumulated perimeter of openings does not exceed half of the circle, otherwise it can not be guaranteed that the solution is unique. The circular averages, that are necessary to calculate the non-harmonic residual according to Equation 4.15, are obtained directly from the border-repairing algorithm as the zero-order Fourier coefficient in the end-point spectrum.

4.2.3   Expeditious calculation of the monopolar term in the 2D Laplacian

It is considered that the average value of the 2D spatially-unwrapped GRE phase $\varphi^{\text{exp}}(r, \Omega)$ along a circle of given radius R is:

$$\langle \varphi^{\text{exp}}(R) \rangle = F(R) = \frac{\int_0^{2\pi} \varphi^{\text{exp}}(R, \Theta) \cdot R \cdot d\Theta}{\int_0^{2\pi} R \cdot d\Theta} = \frac{1}{2\pi} \int_0^{2\pi} \varphi^{\text{exp}}(R, \Theta) d\Theta \qquad (4.23)$$

This average value along a fixed-center circle depends only on the radial coordinate R because the angular coordinate Θ is consumed by the fixed-limits integration. The functions with constant Laplacian in 2D, acting as monopolar non-harmonic terms of the GRE phase map, are x^2 and y^2 or any linear combination of them.

Following the behavior of the harmonic function $F(R)$ is described for $\varphi^{\text{exp}}(R, \Theta) = a \cdot x^2 = a \cdot r^2 \cdot cos^2\Theta$, where a is an arbitrary real number:

$$F(R) = \frac{1}{2\pi} \int_0^{2\pi} a \cdot R^2 \cdot cos^2\Theta \cdot d\Theta \qquad (4.24)$$

$$= \frac{a \cdot R^2}{2\pi} \int_0^{2\pi} \frac{1 + cos(2 \cdot \Theta)}{2} \cdot d\Theta \qquad (4.25)$$

$$= \frac{a \cdot R^2}{2\pi} \cdot \pi \qquad (4.26)$$

$$= \frac{a \cdot R^2}{2} \qquad (4.27)$$

The harmonic part of the 2D phase map (i.e. the main component) will add a constant value to $F(R)$ independent of R. This is the fundamental average property of harmonic functions on a sphere (here a circle in the 2D restriction). Therefore, it has to be considered that:

$$F(R) = \frac{a \cdot R^2}{2} + C \qquad (4.28)$$

To eliminate the constant C and to isolate the influence of the non-harmonic term $a \cdot x^2$, a subtraction of the two values of $F(R)$ at different radii is performed. For instance R and $R + \Delta R$ can be used, where ΔR is the scalar increment of the radius:

$$F(R + \Delta R) - F(R) = \frac{a}{2} \cdot \left[(R + \Delta R)^2 - R^2 \right] \qquad (4.29)$$

$$= \frac{a \cdot \Delta R}{2} \cdot (2 \cdot R + \Delta R) \qquad (4.30)$$

Further on: $\nabla^2(a \cdot x^2) = 2 \cdot a = \epsilon = const$ or, equivalently, $\epsilon = 2 \cdot a$, where equation 4.13 is recalled. ϵ is the spatial average of the residual 2D Laplacian (e.g. zero order approximation) of the GRE phase map. Given that a is the same constant in equation 4.30 and equation 4.13, the fundamental relationship of interest is:

$$F(R + \Delta R) - F(R) \ = \ \frac{\epsilon}{4} \cdot \Delta R \cdot (2 \cdot R + \Delta R) \qquad (4.31)$$

The same relationship can be demonstrated for a non-harmonic term in the GRE phase map of the type y^2, and also for any linear combination of the type $a \cdot x^2 + b \cdot y^2$ with a and b arbitrary real numbers. The equation 4.31 is equivalent to the equation 4.15, q.e.d.

4.2.4 Influence of gradient-induced eddy currents on the GRE phase harmonicity

The reference-less approach developed and discussed in this thesis based on the condition that eddy currents can be neglected. If not-neglectable eddy currents occur, they will induce non-harmonic terms in the 3D GRE phase map and will have potential impact on equation 4.12. Such eddy currents are known to be induced in electrically conductive medium by the switching of imaging gradients, hence producing theoretically rotational magnetic field perturbations. The general expression for the magnetic field produced directly by the imaging gradient coils is written as:

$$\vec{B}^{\text{grad}} = \left(0, 0, \vec{G} \cdot \vec{r}\right) \qquad (4.32)$$

The following equations from the fundamental Maxwell's system will be used:

$$\nabla \times \vec{E} \ = \ -\frac{\partial \vec{B}^{\text{grad}}}{\partial t} = \left(0, 0, -\frac{\partial \left(\vec{G} \cdot \vec{r}\right)}{\partial t}\right) \qquad (4.33)$$

$$\vec{j_c} \ = \ \sigma_e \cdot \vec{E} \qquad (4.34)$$

where $\vec{j_c}$ is the free current density, $\vec{E}$ is the electrical field dynamically induced in matter because of magnetic gradients switching and σ_e is the electrical conductivity in biological tissue.

Substitution of the phenomenological expression of the eddy-current density in the right term of the Equation 4.1 leads to:

$$\vec{\nabla} \times \vec{H}^{\text{eddy}} = \vec{j_c} = \sigma_e \cdot \vec{E} \qquad (4.35)$$

Combining equations 4.33 and 4.35:

$$\vec{\nabla} \times \vec{\nabla} \times \vec{H}^{\text{eddy}} = \sigma_e \cdot \vec{\nabla} \times \vec{E} = \left(0, 0, -\sigma_e \cdot \frac{\partial \left(\vec{G} \cdot \vec{r}\right)}{\partial t}\right) \qquad (4.36)$$

Using standard operator calculation:

$$\vec{\nabla} \times \vec{\nabla} \times \vec{H}^{\text{eddy}} = \vec{\nabla}\left(\vec{\nabla}\vec{H}^{\text{eddy}}\right) - \vec{\nabla}^2\vec{H}^{\text{eddy}} = -\vec{\nabla}^2\vec{H}^{\text{eddy}} \qquad (4.37)$$

where $\vec{\nabla}\vec{H}^{\text{eddy}} = 0$ as the magnetic intensity has zero divergence in a homogeneous medium. By combining equations 4.36 and 4.37, it is possible to relate the Laplacian of the z-component of the additional perturbation of the local magnetic field (produced via the eddy currents induction) to the dynamic variation of magnetic gradients as:

$$\vec{\nabla}^2\vec{H}_z^{\text{eddy}}(\vec{r},t) = \sigma_e \cdot \frac{\partial\left(\vec{G}\cdot\vec{r}\right)}{\partial t} \qquad (4.38)$$

This additional z-component in the magnetic field will modify the local Larmor frequency and will yield an additional phase of spin rotation in the transversal plane. Considering the time interval from the origin of the excitation pulse ($t = 0$) to the center of the gradient-recalled echo ($t = TE$) the additional phase is given by:

$$\varphi^{\text{eddy}}(\vec{r},t) = \mu_0 \cdot \gamma \cdot \int_0^{TE} H_z^{\text{eddy}}(\vec{r},t)\cdot dt \qquad (4.39)$$

Combining equations 4.38 and 4.39, the contribution of eddy-currents to the Laplacian of the GRE phase is given by:

$$\begin{aligned}
\vec{\nabla}^2\varphi^{\text{eddy}}(\vec{r},t) &= \mu_0\cdot\gamma\cdot\int_0^{TE}\vec{\nabla}^2 H_z^{\text{eddy}}(\vec{r},t)\cdot dt &(4.40)\\
&= \mu_0\cdot\gamma\cdot\sigma_e\cdot\int_0^{TE}\frac{\partial\left(\vec{G}\cdot\vec{r}\right)}{\partial t}\cdot dt &(4.41)\\
&= \mu_0\cdot\gamma\cdot\sigma_e\cdot\left[\left(\vec{G}\cdot\vec{r}\right)|_{t=TE} - \left(\vec{G}\cdot\vec{r}\right)|_{t=0}\right] &(4.42)
\end{aligned}$$

Equation 4.42 includes some fundamental physical constants, tissue dependent electrical conductivity, spatial coordinates and gradient strengths. To evaluate the order of magnitude of the residual Laplacian due to eddy currents, following estimates are used: $\sigma_e \approx 0.1\,\text{S/m}$ in liver [44], $r \approx 0.1\,\text{m}$ as a typical order of distance, and $G \approx 30\,\text{mT/m}$ as a typical value of maximum gradient strength in clinical MR systems. Under these conditions:

$$\vec{\nabla}^2\varphi^{\text{eddy}} \leq 0.1\frac{\text{rad}}{\text{m}^2} \approx 10^{-5}\frac{\text{rad}}{\text{mm}^2} \qquad (4.43)$$

or equivalent:

$$\epsilon^{\text{eddy}} \leq 10^{-5}\frac{\text{rad}}{\text{mm}^2} \qquad (4.44)$$

This contribution to the ϵ-parameter originating from eddy currents is 2 orders of magnitude smaller than any significant value of ϵ found experimentally due to the 2D restriction of the geometry (see Table 4.3 and 4.4), and also smaller than the experimental error in calculating the ϵ-parameter. Therefore, it can be concluded that eddy currents induced in tissue by imaging gradients switching on a clinical system have no measurable effect on the condition of harmonicity satisfied in a homogeneous medium by the GRE phase expressed by the Equation 4.12.

4.3 Material and methods to evaluate the reference-less PRF thermometry

4.3.1 Technical setup and evaluation of pre-clinical examinations

4.3.1.1 Validation of the reference-free thermometry

Two approaches have been used in this study to assess the precision and accuracy of the new reference-free MR thermometry.

Firstly, using static ex vivo tissue heated moderately by HIFU (below the coagulation level), the reference-less temperature calculation was compared with the phase-referenced standard PRF thermometry considered here as the "ground truth". At best, the two calculation methods should provide similar results in this case. One can expect a lower "white" noise level in reference-less temperature maps as the calculated background is a smooth function. The agreement between conditions in heated voxels was quantified by Lin's method [76,77]. The level of agreement was assessed using the calculated concordance correlation coefficient (CCC) classified as excellent (0.81-1.00), substantial (0.61-0.80), moderate (0.41-0.60), fair (0.21-0.40), slight (0.00-0.20) or poor (< 0.00). Next, Bland and Altman [5] analysis was applied to further assess the relative contribution of the bias and error for the differences between conditions. The hypothesis of zero/no bias was tested by a paired t-test. All statistical analysis was carried out using Stata 10 (College Station, TX) statistical software. In unheated voxels distant from sonication focus/foci (i.e. baseline measurement), only the Bland and Altman analysis was applied. The effect of noise is attenuated in the reference-free reconstructed background due to the averaging feature of the iterative convolution.

Secondly, baseline measurements were performed in the liver of healthy volunteers (written informed consent). At best, the calculated background phase inside the ROI should be similar in this case to the experimentally measured phase (after spatial unwrapping); only differing by the intrinsic noise of the measurement since the reconstructed phase is smooth.

No external sensor for temperature measurement was used in this study (e.g. thermocouple or fluoroptic fiber) because such a device could disturb the local magnetic susceptibility, may contaminate with air bubbles the tissue around the tip during insertion and acts as a strong reflector for the HIFU beam thus inducing strong alterations in the thermal pattern and increasing the partial volume effects at the interface with the tissue. For standard PRF thermometry, the reference phase map was defined as the temporal average of five pre-sonication measurements in order to decrease the average "white" noise by a factor of $\sqrt{5}$. Zero-order B_0-drift correction was implemented by subtracting the spatial average phase from an unheated user-defined ROI (including 50 to 70 pixels). Temporal phase unwrapping was performed iteratively; assuming the temperature-induced phase shift does not exceed π in the temporal window from dynamic scan n to dynamic scan $n + 1$.

For reference-free thermometry, the definition of the unheated border depended on the heating pattern expected for a given slice orientation. When performing single focus sonication, a closed circular border was defined for the plane orthogonal to the beam and a symmetrical double opening circular border (as shown in Figure 4.3 a) is suggested to be used for the plane parallel to the beam. When performing volumetric sonication, the temperature elevation in the plane orthogonal to the beam was calculated independently for two single-opening circular borders (see Figure 3 for details) and was further fused into a unique temperature map. In the areas where the two domains overlap, the average value of the two independent calculations for the pixel was used.

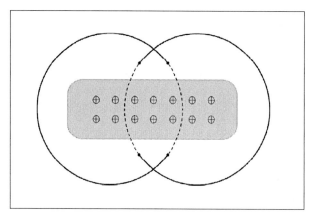

Figure 4.4: *Example of a double open-border implementation of the current reference-free method for MR thermometry, appropriate for monitoring volumetric HIFU sonication (see 1.2.3). The background phase information along the dotted line segments of the border is reconstructed as explained in Figure 4.3, as they cross the heated zone. After border reconstruction, the background phase inside each domain is calculated as illustrated in Figure 4.2 and combined into the final phase background.*

4.3.1.2   Experimental setup of the pre-clinical study

MR thermometry acquisitions were performed on a 3 T whole body MRI scanner (MAGNE-TOM Trio, A Tim system, Siemens Healthcare, Germany). The PRF-sensitive sequence was based on the gradient echo planar imaging (GRE-EPI) kernel with different acquisition parameters depending on the experiment type (see Table 4.1). RF spoiling and lipid suppression were systematically used.

Focused ultrasound heating was produced using a 256-element phased array transducer (Imasonic, 275 Besançon, France) with aperture D = 140 mm, natural focal length R = 130 mm and nominal frequency range from 968 kHz to 1049 kHz. The phased array was driven by a programmable 256-channel generator (Image Guided Therapy, Pessac-Bordeaux, France) with independent control of amplitude and phase per channel and was moved in the horizontal (x_{Oz}) plane by a 2D positioning system with piezoelectric actuators (Image Guided Therapy, Pessac-Bordeaux, France). Ex vivo sonications were performed in samples of turkey muscle degassed for 30 minutes under vacuum. Experimental simulation of the respiratory motion was obtained by using an inflating balloon driven by a mechanical ventilator, as shown in Figure 4.5. The balloon was circumscribed by 5 faces of a parallelized support, while in the direction of the missing 6th face of the support the balloon was in contact with an US-transparent bag containing degassed ex vivo liver tissue. The amplitude of the periodic motion while inflating/deflating the balloon ranged from 1 to 3 cm depending on the ventilator volume set by the driver. The period of motion was adjusted in the range of 4 to 7 sec/cycle. An appropriate acoustic window from the HIFU transducer enabled the ex vivo tissue to be sonicated, as visible in Figure 4.5.

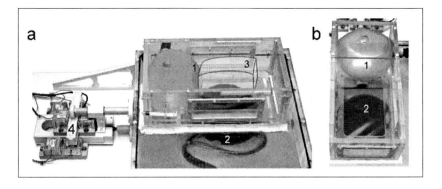

Figure 4.5: *Ex vivo simulator of breathing motion during MR guided HIFU therapy (see 1.2.3). An inflating balloon (1) is driven by a mechanical ventilator and induces motion of the ex-vivo specimen (3). The HIFU transducer (2) is positioned by means of a 2D piezo-electric motorized system (4).*

Four independent single-point sonications at spared locations were performed in the liver of one sheep (female, 35 kg) in vivo, with protocol approval by Geneva University Institutional Animal Care and Use Committee and by the Cantonal Veterinary Authority of Geneva. The animal was pre-medicated with ketamine (30 mg/kg; Pfizer, Zurich, Switzerland) and midazolam (0.2 mg/kg; Roche Pharma, Reinach, Switzerland). After placing an intravenous access using a jugular vein, the sheep was intubated and mechanically ventilated (Servo Ventilator 900D, Siemens-Elema, Sweden) using 6 m long ventilation tubings, and anesthesia was maintained by continuous inhalation of 2% isoflurane (Abbott AG, Baar, Switzerland). The

Table 4.1: Main parameters of the PRF-sensitive sequences used in this study:

Purpose	FOV [mm]	Number of slices	Voxel size [mm]	TR TE ES [ms]	EPI factor PF Parallel imaging	Flip angle Band width [Hz/Px]	Lipid surp. Flow comp.	Acquisition time per Volume [s]	Coil
Ex vivo static	128x128	5 (3 cor, 1 sag, 1 tra)	1x1x5	184 8.5 to 12 2.2	9 6/8 no	10° 528	FATSAT no	3.7	surface loop Φ=11cm
Ex vivo motion simulator	128x128	1 cor	1x1x5	184 8.5 to 12 2.2	9 6/8 no	10° 528	FATSAT no	0.6	Body matrix (2 elements) (2 elements)
in vivo r.g.	128x128	3 (1 cor, 1 sag, 1 tra)	1x1x5	184 8.5 to 12 2.2	9 6/8 no	15° 528	FATSAT no	2.7	surface loop Φ=11cm
Free breathing volunteer baseline	300x300	1 (sag or cor)	2.6x2.3x8	23.4 to 29 9.7 to 9.9 3.5	7 6/8 mGrappa=2	15° cor: 930 sag: 797	"121" pulse yes	0.3	Body matrix (2 elements) + Spine array (3 elements)

r.g. = respiratory triggered, ES = echo spacing, BW = band width, PF = partial Fourier, Lipid surp. = Lipid suppression, Flow comp. = Flow compensation, cor = coronal, sag = sagittal, tra = transversal breathing rate was approximately 9-11 breaths/min. Blood oxygen saturation, body temperature and exhaled CO_2 were monitored continuously. Immediately prior to HIFU sonication, an i.v. injection of pancuronium (Essex Chemie AG, Lucerne, Switzerland) quell dosage was administrated to prevent any accidental muscular contraction during MR guided HIFU ablation, hence ensuring a regular respiratory motion and accurate trigger/gating. An optical sensor encoded the expansion of the abdomen during breathing and the modulated signal was used to trigger the MR thermometry and to gate the HIFU sonication during the quiet phase of respiration. The overall duty cycle of sonication was approximately 40%. The applied peak power was 240 acoustic W. The sonication window covered 5 respiration cycles (approximately 30 s). The delivered acoustic energy was approximately 2800 J per independent sonication.

4.3.1.3   Acquisition of thermometry data in volunteers

Baseline acquisition was performed on five healthy volunteers under free breathing, upon signed written agreement (4 male, 1 female, age range 25 to 35 years). Each volunteer was positioned in the supine position and dynamic acquisition of single-slice thermometry data was performed for 30 s to 60 s (see Table 4.1 for details on sequence), corresponding to 100 to 200 dynamic measurements. The slice orientation was set to sagittal, and then coronal, in independent acquisitions. The precision and the accuracy of the near-harmonic reconstruction of the background phase were evaluated for different sizes and positions of the circular ROI, see Figure 5. The read-out flow compensation option was available for the sequence used on the volunteers.

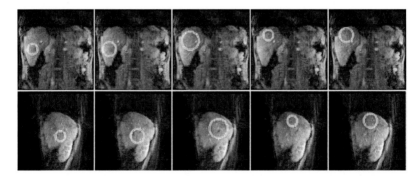

Figure 4.6: *Illustration of the different sizes and positions of the circular ROI used for near-harmonic reconstruction of the background phase in healthy volunteer liver. See Table I for the acquisition parameters of the temperature sensitive sequence.* **Upper row**: *coronal plane, end of expirium.* **Lower row**: *sagittal plane, end of inspirium. The double layer border (necessary to calculate the non-harmonic term of the phase map) is explicitly shown. From left to right the position and diameter (D) of the inner circle are, respectively: lower segments (D = 33.6 mm), lower segments (D = 48 mm), central (D = 62.4 mm), sub-diaphragmatic (D = 33.6 mm), sub-diaphragmatic (D = 48 mm).*

4.3.2 Technical setup and evaluation of first patient data

4.3.2.1 Clinical protocol for LITT ablation in liver

Thermal ablations of liver malignancies were performed on a 1.5 T MR clinical scanner (Magnetom Avanto, Siemens Healthcare, Erlangen, Germany) using LITT in 8 patients. The study protocol was approved by the local ethics committee and written informed consent was acquired from all patients. Prior to the intervention, patients received a local anesthesia (20 ml of 1% prilocaine subcutaneously) and a mild anxiolytic i.v. (Haloperidol: 10 mg haloperidol, 100 mg pethinide; TEVA Generics GmbH, Radebeul, Germany). All procedures were performed by an experienced interventional radiologist. A Neodym YAG-Laser (Neodym doped Yttrium-Aluminium-Granat-Laser, Medilas fibertom 5100, Dornier, Wessling, Germany) was used, operating at a wavelength of 1064 nm and a maximum power of 100 W, which can irradiate with up to 3 applicators simultaneously. For the interventions, a microcatheter system, with an outer diameter of 1.8 mm (5.5F) (Anacath, Berlin, Germany), was placed under MR guidance. After insertion, each titanium needle was replaced with a laser light guide (diffusive tip length = 3 mm) and thermotherapy was initiated.

Table 4.2: *Overview of clinical data for the patients included in the study.*

Patient	Tumor size (cm) transversal	Tumor size (cm) sagittal	Liver segment	Number of applicators	Laser energy (kJ)	Tumor Histology
1	3.8x3.3	3.5x2.7	8	3	48.1	GC
2	2.2x2.1	2.5x2.2	4	2	32	RC
3	5.0x2.7	4.9x3.3	8	3	37	GC
4	2.0x1.9	2.2x2.1	8	2	32.7	CM
5	5.0x3.8	4.8x3.1	2	3	44	RC
6	5.2x4.8	4.8x3.7	4	2	27.5	BC
7	0.9x1.1	1.0x1.2	7	1	17.3	CM
8	1.9x0.6	1.7x0.7	4	1	16	SC

LITT ablation was monitored on line using the standard reference method (Patient 1-6) and, the reference-less method (Patient 7-8).
GC=gastric adenocarcinoma, RC=rectum carcinoma, CM=colorectal metastases, BC=breast carcinoma, SC=sigma carcinoma

On-line temperature monitoring simultaneous to the LITT was initially implemented with a GRE sequence (used for the first six patients), and later a faster segmented-EPI GRE was used for the two additional patients under a similar protocol. The GRE sequence (TA/TR/TE = 950 ms/22 ms/12 ms; $\alpha = 35°$; BW = 250 Hz/Px; voxel = $2.5x2.5x6\,mm^3$; FATSAT lipid signal suppression, FOV = $320x320\,mm^2$, matrix size: 128x128, parallel acquisition GRAPPA = 3, body array coil) was triggered using the scanner's built-in breathing sensor belt. In each breathing cycle, three transversal slices and one sagittal slice were acquired sequentially. Temporal resolution depended directly on the breathing cycle intervals of the patient. The segmented-EPI GRE sequence (TA/TR/TE = 300 ms/22 ms/12 ms;

$\alpha = 15°$; BW $= 910\,\mathrm{Hz/Px}$; voxel $= 2.5x2.5x5\,mm^3$; EPI $-$ factor $= 13$; "121" water excitation; FOV $= 320x320\,mm^2$, matrix size: $128x128$, parallel acquisition GRAPPA $= 2$, 6 element body array coil plus 6 elements from the spine array coil) was used to enable free breathing and high temporal resolution for online temperature monitoring without the requirement for respiratory triggering. One slice was acquired in $300\,\mathrm{ms}$. Three orthogonal slices (sagittal, coronal, transversal) were sequentially acquired in order to deliver a pseudo 3D-mapping of temperature distribution with a temporal resolution of approximately $1\,\mathrm{s}$.

For all interventions, the same power level ($15\,\mathrm{W}$ per applicator) and cooling scheme were prescribed for 20 minutes. The effective delivered energy was recorded by the device, as listed in Table 4.2. During the entire intervention, on-line temperature monitoring was used as a safety tool, to ensure that no at risk organs or vessels were exposed to the laser heating. Thus the temperature elevation was monitored on-line for each treatment and displayed on the main screen of the MR computer, as a color overlay to the corresponding magnitude image. The underlying method of automatic temperature calculation behind the on-line display was the standard reference subtraction in the first 6 patients (monitored with the GRE sequence) and the near-harmonic 2D reference-less method in the last 2 patients (monitored with the segmented EPI GRE sequence). Computing time for reference-less MRT was less than 100 ms/slice. For the standard reference thermometry, the radiologist had to acquire a multi-slice reference image before heating and to define slice-specific ROIs where the calculated relative temperature values were visualized. For the reference-less thermometry, the radiologist had to position the slice-specific ROI for the background reference phase calculation and could define one or two open segments in the border to exclude possible heating or artifacts. The radius of this ROI for the reference-less MRT calculation ranged between 10 and 17 voxels (5 to 8.5 cm diameter). Note that in both scenarios, user interaction was only required at the beginning of the intervention to position the monitoring ROI (standard reference method) or to define the border for the background phase calculation (reference-less method). To control the positioning of the opened border (required for reference-less background reconstruction), the border was visualized within the online display. In this study, a single open segment of the border was defined in each slice, aiming to avoid border phase conflicts such as heating, flow artifacts originating from vena cava, or susceptibility discontinuities whenever malignant tissue lied near the liver capsule. Length and angular position of the open segment were defined by two parameters, with a maximum size restricted to a quarter of a circle. The first parameter was the angular origin of the segment along the circle, while the second parameter defined the ratio between the length of the open segment and the total perimeter of the circle (expressed as a percentage), see Figure 4.7. Once configured, the same border was used until the end of the ablative intervention.

The follow-up examination (24h after intervention) included bolus contrast-enhanced (0.2 mL/kgBW i.v. Gadovist, Bayer-Schering, Berlin, Germany) MR-images (FLASH 2D sequence with a TR = 115 ms, TE = 5 ms, TA = 1 s, 30 slices, $\alpha = 70°$, no fat suppression, same voxel size as for the thermometry sequences). Unperfused tissue was defined as non-enhanced (hypointense) regions after the administration of the contrast agent.

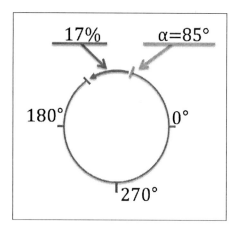

Figure 4.7: *Open border with one open segment defined using two parameters: a starting angle α and the ratio of the segment length to the circle perimeter.*

4.3.2.2   Retrospective evaluation of data after intervention

The raw data from the thermometry acquisition for each patient was extracted from the host computer of the scanner and post-processed offline on an external PC, using the same image calculation environment (ICE, Siemens Healthcare, Erlangen, Germany) as was running in real time on the MR image reconstruction CPU.

For the first 6 patients (that were scanned with the respiratory gated MRT sequence) an explicit comparison was performed between the standard reference method and the novel reference-less method. The two methods were applied to identical raw data sets on a slice-per-slice basis. The relative tissue temperature was calculated once in an unheated liver region (i.e. baseline assessment, or "zero" measurement stability) and once in a LITT-heated ROI.

The unheated ROI was a circle of radius 3 pixels positioned within the liver but sufficiently far from the LITT applicator(s) as to evaluate the precision of each MRT method (standard reference and near-harmonic reference-less). The spatio-temporal stability (or precision) of the MRT baseline was calculated using a two-step process. Firstly, a temporal standard deviation of temperature was determined on a pixel-by-pixel basis from the series of dynamic measurements over the period of the LITT energy application. That is, a map of pixel-by-pixel SD was established. Secondly, a mean value of the SD in the unheated ROI was calculated over space and averaged over all slices, yielding a global spatio-temporal indicator of MRT precision. It must be taken into account that for the standard reference method, a B_0-correction must be performed to correct for the phase drift [106], visible in Figure 4.8. This phase drift was removed before computing the SD. For the reference-less method, no B_0-correction is necessary as any B_0-drift components are automatically removed in the calculation process.

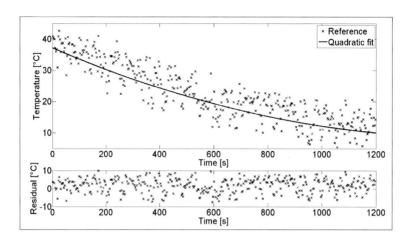

Figure 4.8: *Temperature vs. time curve in a voxel from one unheated ROI in the liver of patient 4 (see table 4.2) calculated using the standard reference-subtraction method, applied to respiratory triggered acquisition under free breathing.* **Upper:** *The magnetic field drift over time leads to $30\,°C$ apparent temperature variation over 20 minutes, which has to be corrected for. For illustration, the drift was fitted as a second order polynomial with coefficients of order 0, 1 and 2 equal respectively to [37.2, 0.037, 0.000012].* **Lower:** *The residual standard deviation between the fitted curve and the measured data is $3.9\,°C$ and estimates the intrinsic noise of the MRT acquisition.*

As the last two patients in the series were scanned with a rapid MRT sequence under free breathing and without respiratory triggering (see Table 4.2), no direct comparison of temperature calculation methods was performed in post processing. The performance of the reference-less MRT method presented in this thesis was assessed for this data from the spatio-temporal stability of the baseline in unheated regions (i.e. "zero" measurement stability), see above.

An indirect evaluation of the measured reference-less temperature elevation was also implemented. After the raw-data was post-processed and the reference-less T-map was calculated, the data was fitted over space using a 2D-Gaussian shape for each slice and each acquisition. Furthermore, the difference map between the Gaussian shape and the reference-less temperature data was calculated on a slice-per-slice basis. Finally, for each slice the median of the difference map was calculated within the heated ROI for each acquisition.

The theoretical SD of the MR thermometry was evaluated from the magnitude SNR values [22, 123]. The magnitude SNR was calculated in the same unheated ROI of the liver as defined above. The SNR of the magnitude images was calculated using the dual acquisition subtraction method [34, 42] (section 2.2.3):

$$\text{SNR} = \sqrt{2}\frac{S_1}{\text{SD}_{1-2}} \tag{4.45}$$

where S_1 is the mean signal intensity of a non-heated ROI in the first acquired magnitude image in the liver, and SD_{1-2} is the standard deviation in the same ROI in a difference image. The difference image was calculated by subtracting two consecutive GRE magnitude images. For each measurement, the SNR was averaged over the acquired 4 slices. The theoretical SD (STD), assuming that white noise is the only source of error, was calculated according to Salomir et al. [123] and Conturo et al. [22] as:

$$STD \cong \frac{400}{B_0 \cdot TE \cdot SNR} \tag{4.46}$$

4.3.2.3   Model-based temporal regularization of the temperature data

An additional goal of this thesis consisted in evaluating the potential benefit of coupling the reference-less thermometry with a model-based regularization, performing as an fast temporal filtering, aiming to further improve the thermal monitoring (Section 1.4.1). This approach is meaningful as LITT heating is a slow process compared to the temporal resolution of the MRT. A thermal physical model was used to fit, the measured temperature vs. time curves in each voxel. The idea is to improve the information of the last sampling point by considering the fit-predicted value (instead of the measured one), based on the past information of all previous sampling points acquired since the beginning of the treatment. This expanding-window process was applied to each newly acquired time-point. The number of sampling points included in the fit therefore linearly increases with time. This model-based temporal regularization is compatible with real-time implementation, as the prediction is calculated for each most recent sampling point. This approach has been shown to reduce the influence of artifacts occurring over short time intervals (e.g. for a few sampling points, see 3.2.3.1) when using the standard reference PRF thermometry. The same approach was applied to the data post-processed with the reference-less thermometry method.

The model-constrained based regularization was performed off-line using MATLAB (Mathworks, Natick, MA, USA version 6.0). The underlying physical model was derived from the 3D solution of the Pennes' Bioheat equation [16, 104] written for homogeneous tissue. The temperature distribution over time during heating was fitted for each voxel and for each new measurement using the following relation (see 3.2.3.1):

$$T(r, z, t) = A \cdot t^{-\frac{3}{2}} \cdot exp(\frac{-r^2}{4\kappa t}) \tag{4.47}$$

where κ is the thermal diffusion coefficient, r is the distance from the heating source, t the time, and A is scaled by the diffusion.

The mean SD of the residual differences between the end-point fit curve (i.e. when all sampling points of the heating period are available) and the experimental data (standard reference and reference-less technique, independently) were calculated in order to evaluate the precision and accuracy of the standard reference and, respectively, the near-harmonic 2D reference-less method. Moreover, the SD was calculated for the residues between the end-point fit curve and the regularized data.

Analyzing the SD was a two-step process. Firstly, a pixel-wise SD between each two temporal curves was calculated for each voxel within the ROI, resulting in a SD-map characterizing the whole heating period. Secondly, for each patient, a mean value of the SD-map was calculated, including all voxels within the ROIs from all slices.

4.3.2.4 Thermal dose calculation and prediction of ablation regions

The model-based regularized values of temperature elevation (for the reference-less technique only) were used to calculate the delivered thermal dose. The thermal dose was calculated based on the model described by the **Arrhenius damage integral** [117] (Section 1.4.2) using the same Arrhenius parameter as described in section 3.2.3.2. To compare the unperfused area in the follow up images (24h after intervention) against the calculated thermal dose, the maximum and minimum diameters along two orthogonal directions were estimated in the sagittal and transversal planes. Among the three transversal slices of the MRT, the one with the largest lethal dose extension was chosen. Bland and Altman [5] analysis was applied to assess the relative contribution of the bias and error for the differences between conditions, using Stata 10 (College Station, TX) statistical software.

4.4 Results of the reference-less PRF method evaluation

4.4.1 Results of the pre-clinical study

4.4.1.1 Ex vivo performance under moderate HIFU heating

Figure 4.9 illustrates the pixel-by-pixel comparison between the calculated temperature in ex vivo static tissue under fixed HIFU focus (moderate heating) using the standard PRF temporal subtraction and using the new reference-free method.

- At the focus (black curve in Figure 4.9.d) statistical parameters for standard subtraction versus near harmonic reference-free calculation were found as CCC: 0.998, 95% confidence interval (CI): [0.997; 1.000], Bland and Altman difference (standard PRF versus reference-less): $-0.082 \pm 0.0335\,^\circ$C, 95% Limits Of Agreement (LOA): [-0.738; 0.575]$\,^\circ$C, Pearson's r = 0.999.

- In a pixel at a distance of 3 mm from the focus (green curve in Figure 4.9.d) CCC: 0.993, 95% CI: [0.987; 0.998], Bland and Altman difference (standard PRF versus reference-less): $-0.069 \pm 0.343\,^\circ$C, 95% LOA: [-0.742; 0.604]$\,^\circ$C, Pearson's r = 0.996.

- In a pixel at a distance 11 mm from the focus (thus unheated) (blue curve in Figure 4.9.d) Bland and Altman difference (standard PRF versus reference-less): $-0.284 \pm 0.284\,^\circ$C, 95% LOA: [-0.840; 0.273]$\,^\circ$C.

Figure 4.9: *Ex-vivo comparison between standard PRF thermometry (B_0-drift corrected), versus the reference-free calculation using the near-harmonic reconstruction of the background phase, for 8 s fixed focus sonication. The imaging plane was orthogonal to the HIFU beam axis (shown FOV is 40 mm square). The implementation of the reference-free calculation used a closed circular border of inner radius 15 pixels (15 mm). **a).** coronal temperature maps temporally integrating the sonication window 04 s; **b).** coronal temperature maps temporally integrating the sonication window 48 s; **c).** coronal temperature maps temporally integrating the post-sonication window 48 s. Temperature elevation color bar is gridded in °C; **d).** plot of the calculated temperature with time for the two MR thermometry methods, considered at the focus and respectively in a pixel shifted 3 mm and 11 mm from the focus. Statistics for comparison are provided in the body text.*

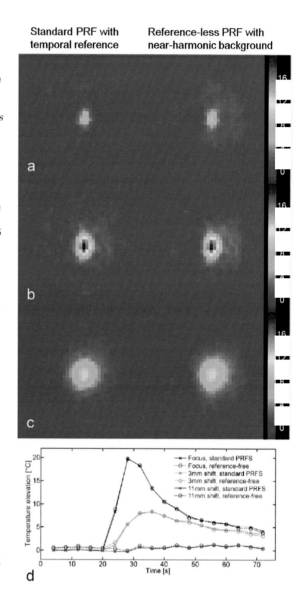

Standard PRF with temporal reference   **Reference-less PRF with near-harmonic background**

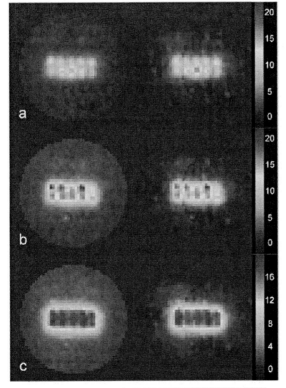

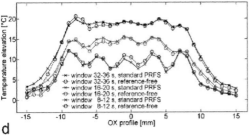

Figure 4.10: Ex-vivo comparison between standard PRF thermometry (B_0-drift corrected), versus the reference-free calculation using the near-harmonic reconstruction of the background phase, for a 10-focus double-line scan volumetric sonication (5 x 2 foci, 4 mm gap, looped at the frame rate of 4 sec as set by the MR acquisition of a data volume). The imaging plane was orthogonal to the HIFU beam axis (shown FOV is 40 mm square). The implementation of the reference-free calculation for this case is explained in Figure 4.4 (each circular border had a radius of 17 pixels and the inter-center distance was 50 pixels). **a).** coronal temperature maps temporally integrating the sonication window 8 12 s; **b).** coronal temperature maps temporally integrating the sonication window 16 20 s; **c).** coronal temperature maps temporally integrating the window 32 36 s. Temperature elevation color bar is gridded in °C; **d).** plot of central spatial profiles through the temperature maps for the two MR thermometry methods, crossing left-right the 5-foci half pattern and considered respectively for the three frames shown in a), b) and c). Statistics for comparison are provided in the body text.

Figure 4.10 illustrates the pixel-by-pixel comparison between the calculated temperature in ex-vivo static tissue under volumetric HIFU sonication (iterating on a 5x2 foci pattern, moderate heating), using the standard PRF temporal subtraction and using the new reference-free method. Statistical parameters of spatial correlation for pixels heated above 5 °C were found to be as follows:

- measurement during sonication window 8-12 s, CCC: 0.951, 95% confidence interval (CI): [0.938; 0.964], Bland and Altman difference (standard PRF versus reference-less):−0.456 ± 0.583 °C, 95% Limits Of Agreement (LOA): [-1.599; 0.687] °C, Pearson's r = 0.972;

- measurement during sonication window 16-20 s, CCC: 0.957, 95% CI: [0.949; 0.966], Bland and Altman difference (standard PRF versus reference-less): 0.968 ± 0.592 °C, 95% LOA: [-0.192; 2.128] °C, Pearson's r = 0.991;

- measurement during sonication window 32-36 s, CCC: 0.976, 95% CI: [0.971; 0.980], Bland and Altman difference (standard PRF versus reference-less): 0.931 ± 0.674 °C, 95% LOA: [-0.39; 2.252] °C, Pearson's r = 0.994;

The advantage of the reference-free method was clearly demonstrated for cases involving tissue motion, as illustrated in Figure 4.11. The mechanical ventilator-driven balloon was inflating/deflating in an approximately binary manner, i.e. high/low level with rapid transition. The reference for the standard PRF was defined as the average phase image over four successive rapid acquisitions during the first high level stage. Therefore it was possible to compare the results of standard PRF thermometry with the new reference-free method during the high level (e.g. inflated balloon) periods. Within the intrinsic noise standard deviation, both methods provided similar values of temperature. Contrarily, during low level periods (e.g. deflated balloon), the standard method yielded large errors, in particular the underestimation of the temperature by several tens of °C, whereas the reference-less method correctly tracked the temperature evolution in the moving tissue, see the stable baseline in Figure 4.11.d.

4.4.1.2   In vivo performance with fixed focus sonication

In the in vivo experiment considered here for illustration, the position of the liver from one respiratory cycle acquisition to another was found to be very similar, as assessed from respiratory triggered magnitude GRE-EPI data in sagittal and coronal planes. The diaphragm position was found reproducible. In particular, no accidental motion of tissue was detected within the spatial resolution available (1 mm). Therefore, the triggered referenced thermometry applied to the forced-breathing acquisition was expected to provide accurate results. As shown in Figure 4.12, the reference-free method provided similar results to the triggered referenced method, improving slightly the baseline stability. The offset between the temperature values provided by the two methods at any given voxel inside the circular ROI was comparable to the intrinsic standard deviation of the acquisition noise.

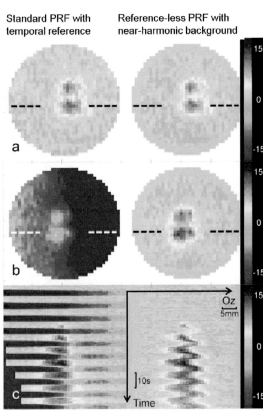

Standard PRF with temporal reference

Reference-less PRF with near-harmonic background

a

b

c

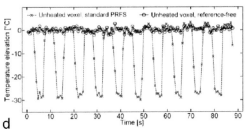

d

Figure 4.11: *Ex-vivo moving phantom comparison between standard PRF thermometry (B_0-drift corrected) and the reference-free calculation using the near-harmonic reconstruction of the background phase. Results are shown for an interleaved double focus HIFU sonication of 40 s (50%-50% duty cycle, 6 mm gap along HF direction). The imaging plane was orthogonal to the HIFU beam axis (shown FOV is 41 mm square). The setup shown in Figure 4.5 has been used, adjusting the ventilation period at 7.5 s. The implementation of the reference-free calculation used a closed circular border of inner radius 16 pixels (16 mm). **a).** coronal temperature maps after 24 s of sonication, inflated balloon stage; **b).** coronal temperature maps after 27.4 s of sonication, deflated balloon stage; **c).** color coded 1D profile of temperature through one focal spot (see dotted line in frames a) and b) scrolled over the time; **d).** plot of the calculated temperature with time, for the two MR thermometry methods, in the same unheated voxel; note the periodic large error of standard PRF calculation and the stable baseline for the reference-free method.*

Figure 4.12: *In-vivo thermometry in sheep liver during fixed focus HIFU sonication.* **a),** **b):** *axial and respectively coronal GRE-EPI magnitude images;* **c), d):** *axial and respectively coronal GRE-EPI phase images; Shown FOV (a-d) is 120 mm square.* **e).** *respiratory triggered standard PRF thermometry (B_0-drift corrected) in coronal plane at the time point of maximum temperature rise;* **f).** *reference-free calculation of temperature using the near-harmonic reconstruction of the background phase, for the same time point. Shown FOV (e-f) is 40 mm square. The implementation of reference-free calculation used a closed circular border of 18 mm diameter;* **g).** *Plots of the calculated temperature with time for the two MR thermometry methods, in one unheated voxel and at the focus. Absolute temperature values are provided. The accuracy and precision of the baseline measurement curves plotted here were $0.59 \pm 1.16\,°C$ for the triggered referenced method and $-0.33 \pm 0.99\,°C$ for the reference-free method.*

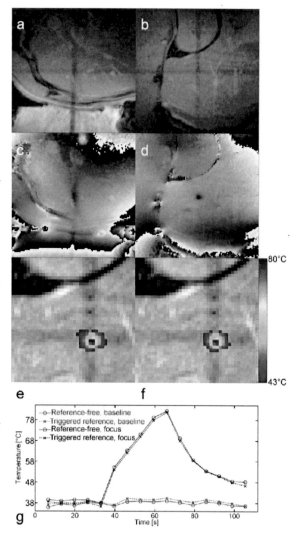

The absolute temperature reached at the focus was approximately 80 °C, the two thermometry methods provided very similar results. Note that under this high temperature heating regime tissue coagulation occurs instantly. The main hypothesis of the current reference-free thermometry is that no sharp variation of magnetic susceptibility distribution exists (according to equation 4.11). The results presented in this thesis show that no significant dynamic change of tissue susceptibility was induced by the rapid coagulation, otherwise the reference-free method used here would have become inaccurate.

4.4.1.3 Stability of reference-less thermometry baseline in unheated humans

The intrinsic noise level of the acquired GRE-EPI phase in healthy volunteers (calculated from a dynamic acquisition during expirium apnea) was found to be 0.16 °C on average, demonstrating excellent stability of the acquisition sequence. The baseline calculation (i.e. zero-measurement) over 94 test cases indicated an average accuracy of 0.13 ± 0.47 °C [range -0.77 to 1.08 °C] and an average precision of 0.65 ± 0.26 °C [range 0.18 to 1.28 °C]. Detailed results are given in Table 4.3 and 4.4. The non-harmonicity of the 2D phase (i.e. local 2D curvature) measured by the ϵ-parameter was found to be of the order of 10^{-4} rad·mm^{-2} to 10^{-3} rad·mm^{-2}, fluctuating both positive and negative, without any evident correlation with the plane orientation or the size of the domain. The general tendency was to obtain an improved precision for smaller domain sizes. No correlation was found between the accuracy and the size of the ROI, nor between the accuracy and the position of the ROI within the liver. The mean precision was found to be improved by 10 to 20% when the ROI was positioned in the lower segments of the liver as opposed to the subdiaphragmatic region for an equal size of domain, however this difference, of the order of 0.1 °C, was not found to be statistically significant.

Table 4.3: *PART A: Precision and accuracy of temperature baseline in the liver of five healthy volunteers for two different slice orientations (see Figure 4.6).*

Volunteer	Slice direction	Respiration phase	Inspirium					Expirium				
		ROI position	Lower segments		Center	Sub-diaphragmatic		Lower segments		Center	Sub-diaphragmatic	
		ROI radius [pixel]	7	10	13	7	10	7	10	13	7	10
		[mm]	16.8	24	31.2	16.8	24	16.8	24	31.2	16.8	24
1 27 year old male	Coronal	$\epsilon[10^{-3}\,\mathrm{rad/mm^2}]$	1.14	1.55	0.63	0.3	0.42	0.56	0.29	0.69	0.28	0.78
		accuracy[°C]	-0.52	-0.62	0.17	-0.09	-0.36	-0.01	0.29	-0.26	-0.07	-0.24
		precision[°C]	±0.38	±0.59	±0.59	±0.58	±0.53	±0.29	±0.41	±0.62	±0.54	±0.63
	Sagittal	$\epsilon[10^{-3}\,\mathrm{rad/mm^2}]$	2.13	1.19	2.38	3.66	3.11	0.76	1.36	2.35	2.57	2.32
		accuracy[°C]	-0.33	0.67	0.72	0.17	0.31	0.42	-0.15	0.27	0.17	0.57
		precision[°C]	±0.26	±0.82	±0.65	±0.46	±0.85	±0.54	±0.81	±0.51	±0.36	±0.50
2 25 year old male	Coronal	$\epsilon[10^{-3}\,\mathrm{rad/mm^2}]$	1.15	0.81	0.64	0.71	0.88	0.70	0.95	0.95	-0.06	-0.09
		accuracy[°C]	-0.16	0.26	0.56	0.13	-0.17	0.35	0.17	-0.07	-0.28	-0.18
		precision[°C]	±0.19	±0.33	±0.81	±0.42	±0.46	±0.26	±0.39	±0.37	±0.44	±0.51
	Sagittal	$\epsilon[10^{-3}\,\mathrm{rad/mm^2}]$	1.92	1.18	2.06	1.99	1.71	2.07	0.93	1.37	1.23	1.03
		accuracy[°C]	-0.01	0.77	0.09	-0.18	0.13	0.23	0.75	0.62	0.11	0.37
		precision[°C]	±0.18	±0.52	±0.67	±0.40	±0.51	±0.34	±0.53	±0.63	±0.51	±0.37
3 35 year old male	Coronal	$\epsilon[10^{-3}\,\mathrm{rad/mm^2}]$	-1.43	NA	NA	-2.17	NA	-1.29	NA	NA	-0.88	-1
		accuracy[°C]	-0.03	NA	NA	0.66	NA	-0.26	NA	NA	0.03	0.08
		precision[°C]	±0.50	NA	NA	±0.96	NA	±0.62	NA	NA	±0.79	±1.19
	Sagittal	$\epsilon[10^{-3}\,\mathrm{rad/mm^2}]$	-1.08	-0.19	0.96	2.52	2.23	2.88	1.15	1.48	-0.22	0.37
		accuracy[°C]	0.61	-0.28	0.96	-0.52	-0.19	-0.77	1.04	-0.27	0.94	0.09
		precision[°C]	±0.52	±0.70	±0.76	±0.52	±0.67	±0.93	±0.89	±1.25	±1.05	±0.95

R = inner radius of the ROI where the background phase is reconstructed as a solution to Dirichlet's problem expressed in pixels (1 pixel = 2.3 mm); ϵ = average curvature (i.e. residual constant Laplacian) of the 2D phase map inside each ROI. NA = not available, the liver morphology did not permit to position inside it the respective ROI.

Table 4.4: PART B: *Precision and accuracy of temperature baseline in the liver of five healthy volunteers for two different slice orientations (see Figure 4.6).*

Volunteer	Slice direction	Respiration phase	Inspirium					Expirium				
		ROI position	Lower segments		Center	Sub-diaphragmatic		Lower segments		Center	Sub-diaphragmatic	
		ROI radius pixel	7	10	13	7	10	7	10	13	7	10
		[mm]	16.8	24	31.2	16.8	24	16.8	24	31.2	16.8	24
4 28 year old female	Coronal	$\epsilon[10^{-3}\,rad/mm^2]$	1.15	1.65	0.18	-1.32	-0.27	0.89	1.00	0.60	-0.99	-0.86
		accuracy[°C]	0.17	-0.73	0.20	0.40	0.01	-0.05	-0.22	0.50	-0.24	0.10
		precision[°C]	±0.35	±0.81	±1.02	±0.46	±0.71	±0.56	±0.25	±1.13	±0.84	±1.07
	Sagittal	$e\,[10\text{-}3\,rad/mm2]$	-0.43	NA	-1.38	-1.58	-0.78	1.07	NA	NA	-1.23	-0.78
		accuracy[°CC]	-0.52	NA	0.59	0.42	-0.68	0.31	NA	NA	0.43	0.17
		precision[°C]	±0.97	NA	±1.17	±0.62	±0.99	±0.86	NA	NA	±0.55	±0.77
5 28 year old male	Coronal	$\epsilon[10^{-3}\,rad/mm^2]$	1.11	0.39	1.15	2.37	1.63	-0.37	0.26	1.15	0.45	1.83
		accuracy[°C]	-0.07	0.30	0.83	-0.16	0.84	0.27	0.50	0.87	0.75	0.50
		precision[°C]	±0.25	±0.49	±0.91	±0.61	±0.72	±0.28	±0.54	±1.02	±0.66	±0.71
	Sagittal	$\epsilon[10^{-3}\,rad/mm^2]$	-0.39	-0.44	-0.94	-0.37	0.39	0.06	0.19	-1.55	-1.65	-1.22
		accuracy[°C]	0.76	0.85	0.31	-0.73	-1.00	-0.37	-0.45	1.08	-0.32	-0.60
		precision[°C]	±0.46	±0.81	±0.1.28	±0.52	±1.12	±0.42	±0.70	±1.00	±0.57	±1.13

R = inner radius of the ROI where the background phase is reconstructed as a solution to Dirichlet's problem expressed in pixels (1 pixel = 2.3 mm); ϵ = average curvature (i.e. residual constant Laplacian) of the 2D phase map inside each ROI. NA = not available, the liver morphology did not permit to position inside it the respective ROI.

4.4.2   Deliverables of first patient data

4.4.2.1   Intrinsic SNR and baseline stability

The 3 pixel radius, circular ROI in the unheated region of the liver was used to evaluate the intrinsic SNR and the accuracy of the thermometry. The calculated SNR ranged from 10 to 17 for the GRE acquisition and was determined to be approximately equal to 13 for the segmented-EPI GRE acquisition (see details in Table 4.5). The SD of the "zero" measurement for the temperature baseline (around the physiologic temperature of 37.2 °C) within the same unheated ROI was found to range between 2.7 °C and 9.0 °C using the standard reference method, and between 1.1 °C and 1.8 °C using the reference-less method (Table 4.5). These latter values were found to be inferior to the theoretical white-noise standard deviation at a given SNR in 7 cases out of 8. This situation is explained by correlations in the measurement noise and also indicates the good quality of the reference-less thermometry near the theoretical limit.

Table 4.5: *Overview of baseline stability results in non-heated ROI's (SD).*

Patient	Non-heated ROI			
	Standard deviation to T=37.2 °C			White noise
	SD	SD	SNR	only theoretical
	Reference [°C]	Reference-less [°C]		STD [°C]
1	3.8	1.7	14±1	1.58
2	2.7	1.1	20±1	1.11
3	9	1.6	15±1	1.48
4	5	1.7	15±1	1.48
5	3.5	1.3	18±1	1.23
6	2.9	1.2	19±1	1.16
7	NA	1.8	14±1	1.58
8	NA	1.2	19±1	1.16

No temporal regularization was performed. All data was averaged over the ROI's in all 4 slices. NA= not available. Theoretical SD (STD) was calculated based on Equation 3.6.

The high level of uncertainty in the temperature baseline using the standard reference method with respiratory triggered GRE acquisition in a clinical scenario in liver indicates that this method alone is not sufficient for deriving a reliable thermal dose and hence a per-operatory end point. The voxel-by-voxel correspondence during the whole intervention was prone to variability under free breathing, despite the respiratory trigger.

4.4.2.2 Calculation of temperature elevation during LITT

The applicator itself did not create any detectable susceptibility artifacts, which could have induced localized errors in the reference-less MRT. The diffusive tip was not visible at the current resolution of the GRE magnitude images before heating.

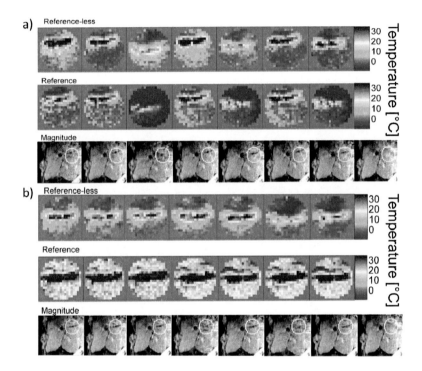

Figure 4.13: *Comparison of 7 consecutive acquisitions in free breathing during LITT of a sagittal plane in patient liver at two different time points,* **a)** *temporal window approximately 4 min from starting the LITT treatment,* **b)** *temporal window after approximately 7 min. Reference-less and standard reference MRT (shown FOV = 70 mm), and corresponding magnitude images (shown FOV = 200 mm) are presented together. The last magnitude image of each series corresponds to the reference scan. Before LITT energy is applied, the optical fiber tip is not visible in the magnitude images.*

The near harmonic reference-less calculation of the temperature baselines showed a significantly higher stability, which may suggest clinical suitability of the method. The benefit of monitoring the temperature with the reference-less calculation instead of the reference subtraction for the respiratory gated GRE acquisition is illustrated in Figure 4.13. A series of

7 consecutive acquisitions (7 breathing cycles) is shown in frame a) in a temporal window centered on $t_1 = 4$ minutes after LITT irradiation started, and another series of temperature maps is shown in frame b) in a temporal window centered on $t_2 = 7$ minutes after LITT irradiation started. The corresponding magnitude images are provided for both cases. The errors visible for the standard reference method increase with time, probably due to slow intra-abdominal drift effects. As expected at the initial stage of the treatment, whenever the liver position was similar between the reference image (shown at right side in Figure 4.13) and the actual acquisition, the standard reference subtraction and the reference-less calculation yielded similar results (shown frame 1,2,4, and 6 in series a)). However, even when the liver position appears to be similar in the reference image and an image at a later stage of heating, the reference-subtraction leads to significant errors (shown frame 3,5, and 7 in series b)).

Table 4.6: *Overview of model-based temporal regularization of temperature data in the heated ROI for the respiratory-triggered MRT acquisition sequence.*

Patient ID	Standard deviation of residues between end-point fitting curve and directly calculated data		Standard deviation of residues between end-point fitting curve and regularized data	
	Reference subtraction [°C]	Reference -less [°C]	Reference subtraction [°C]	Reference -less [°C]
1	16.8 ±8.5	5.3 ±6.5	4.6 ±2.4	1.5 ±2.1
2	6.6 ±2.2	4.4 ±4.4	3.4 ±2.5	2.1 ±1.9
3	14.6 ±6.9	1.8 ±1.9	10.2 ±5.8	0.7 ±1.1
4	8.9 ±8.8	3.1 ±4.6	1.9 ±2.4	0.8 ±1.3
5	8.1 ±3.4	4.4 ±3.1	3.8 ±2.5	2.2 ±1.6
6	7.8 ±3.2	3.7 ±2.3	3.1 ±2.3	1.4 ±1.1

All values were averaged over the ROIs in the 4 slices. (NA= not available).

When comparing the standard deviation of the residues between the directly calculated temperature data and the model-fitting temporal curve, for the two MRT methods (standard reference and near-harmonic reference-less) in the first 6 patients of the protocol, the following ranges were found: from 6.6 °C to 16.8 °C for the case of standard T-mapping and 1.8 °C to 5.3 °C for reference-less T-mapping (see Table 4.6 for details). These results indicate a higher precision for the reference-less method within the heated area by a factor of 3.3 on average. This improvement appeared to be variable from patient to patient, the six individual factors, by order, being approximately: 3.2, 1.5, 8.1, 2.9, 1.8 and 2.1. In general, a greater improvement was obtained in those cases where the initial precision of the standard method was low. The on-the-fly regularization (i.e. expanding-window temporal filtering) based on the BHTE model significantly improved the precision of the temperature estimation in all cases, for both methods (Figure 4.14). As the past information of the slow temporal process was included in the actual temperature prediction, the regularized values of temperature showed decreased spreading around the end-point fitting curve. When the model-based correction was applied to the standard reference MRT data, the standard deviation of the residues of the fitting curve ranged from 1.9 °C to 10 °C, whereas the same correction algorithm working on the reference-less calculation output data leaded to residues of standard deviation ranging from 0.7 °C to 2.1 °C. Note the average ratio of 4 in precision between the two thermometry methods after applying the model-based correction.

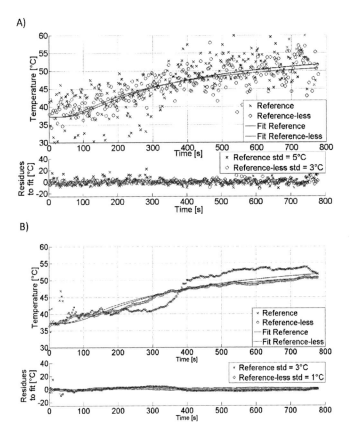

Figure 4.14: *Illustration of pixel-wise calculation of LITT-induced temperature elevation over time in the liver of patient 5. The upper frame (**A**) shows directly calculated values, while the lower frame (**B**) shows the corresponding regularized temperature data. The standard-reference subtraction and the reference-less calculation are both processed. Also, the residues between sampling point data and the end-point fitted curve are plotted below each frame.*

As described previously, the reference-less calculation of the temperature elevation in the last two patients (free breathing, no triggering) was evaluated based on the residues after subtracting a spatial 2D-Gaussian fit (Figures 4.15 and 4.16). This median was plotted over time, an example is shown in Figure 4.16, where the temporal average of the spatial median difference is 0.2 °C.

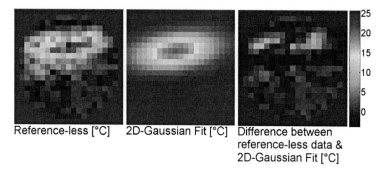

Reference-less [°C]   2D-Gaussian Fit [°C]   Difference between
                                                reference-less data &
                                                2D-Gaussian Fit [°C]

Figure 4.15: *Spatial temperature distribution for one slice in human liver, acquired during heating (free breathing, no respiratory triggers) and calculated using the reference-less PRF method.* **Left** *image shows the temperature map output by the reference-less algorithm,* **middle** *image shows the 2D-Gaussian fit of the same temperature map, and* **right** *image shows the difference between calculated temperature map and the 2D-Gaussian fit. The common color scale of the three frames is graduated in* °C.

4.4.2.3   Spatial correlation between thermal dose and post-treatment unperfused regions

The spatial correlation between the ablation regions (as predicted from the thermal dose calculation with regularized reference-less MRT) and the 24h post treatment unperfused regions, in the first 6 patients, was found to be excellent, see Table 4.7. The Bland and Altman test indicated:

- in the transverse plane: TD/unperfused difference 1.46 ± 2.10 mm with 95% LOA [-2.67; 5.58] mm

- in the sagittal plane: TD/unperfused difference 0.96 ± 2.46 mm with 95% LOA [-3.87; 5.79] mm

Note, the thermal dose calculation showed a minor tendency to overestimate the actual unperfused region, but given the voxel size of 2.5 mm^3 in the plane, these differences are not statistically significant (i.e. average overestimation of lesion diameters similar to half a voxel size).

Table 4.7: *Spatial correlation between the size of the unperfused region (24h post-treatment) and the predicted ablation (thermal dose) from regularized reference-less thermometry data.*

Patient ID	Transverse plane		Sagittal plane	
	Unperfused region at 24h post-treatment [mm] x [mm]	thermal dose region [mm] x [mm]	Unperfused region at 24h post-treatment [mm] x [mm]	term's dose region [mm] x [mm]
1	43 x 43	47.5 x 42.5	41 x 40	45 x 37.5
2	33 x 31	35 x 30	31 x 29	30 x 30
3	49 x 42	50 x 45	48 x 44	50 x 45
4	40 x 35	42.5 x 35	41 x 32	40 x 35
5	48 x 31	47.5 x 30	53 x 31	50 x 32.5
6	50 x 40	52.5 x 45	50 x 41	55 x 42.5

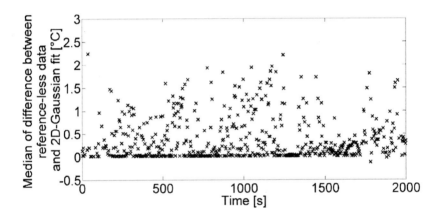

Figure 4.16: *Median of the difference between calculated reference-less data (non-regularized) and their 2D- Gaussian fit [°C] plotted over time, illustrated for patient 7 in the sagittal slice.*

4.5 Advantages and Challenges of the reference-less temperature monitoring

The reference-less MRT method, developed, described and clinically evaluated in this thesis, tries to improve the flexibility, accuracy and mathematical robustness of the temperature calculation in a clinical scenario. The reference-less PRF method presented in this thesis is based on fundamental properties of the static magnetic field inside the matter, and on the theoretical frame of harmonic functions as suggested in reference [124]. Taking a 3D domain with no sources of magnetic susceptibility (or with linearly varying susceptibility), the unwrapped phase of the GRE 3D data is demonstrated to be a harmonic function, which results in a null scalar field when the Laplacian operator is applied. Susceptibility inhomogeneities, localized heating, mixed signal from different chemical shift species with spatially varying ratios, or other artifacts, are all non-harmonic functions which contribute to the globally measured GRE phase. Calculating the background (reference) phase thus means calculating the harmonic part of the phase map. The non-harmonic part of the GRE phase, in a homogeneous medium, constitutes the temperature elevation. Phase information along a one pixel thick border is theoretically sufficient to calculate the full harmonic background phase inside the domain (i.e. inner Dirichlet problem). This border has to be unheated and free from artifacts and susceptibility contrast. To enable slice-per-slice monitoring of temperature, the calculation of the background phase was reduced from 3D to 2D with a residual term in the phase's Laplacian of order zero. This constant Laplacian approximation is called here a near-harmonic function. To determine the residual term of the Laplacian operator, the border has to be at least two pixels thick. For practical reasons (mainly noise-robustness) the border implemented for the study presented in this thesis has a thickness of three pixels. Moreover, openings in the border were compatible with the method by considering a circular geometry of the domain. The open border implementation permits the elimination of areas of conflict from the assigned border, such as heating, artifacts and susceptibility contrast.

The new method for reference-less thermometry based on near-harmonic reconstruction of the GRE phase background developed in this thesis was tested under various conditions. The baseline measurement in healthy volunteer liver was used to evaluate the potential intrinsic errors in the temperature calculation with the new method. The residual errors in the calculated baseline of the temperature include five independent contributions: 1. the intrinsic noise of the measured experimental phase, that is theoretically the minimum error achievable for a given data set, 2. the local deviations from the constant-Laplacian 2D approximation inside a 3D susceptibility-homogeneous medium, including proximal sources of susceptibility from outside the 2D slice, 3. the local sources of magnetic susceptibility in the slice plane and inside the domain, 4. the sub-optimal flow suppression in major vessels yielding to artificial phase shift, and 5. sub-optimal fat signal suppression (when fat is present in the tissue).

Overall, the accuracy and precision of the described method were found to be excellent, and similar to the level of uncertainty due to the intrinsic noise of the measurement. The extensive study [141] on real-time MR temperature mapping in human liver using respiratory gated acquisition on a 1.5 T system indicated a mean temperature SD of 2.3 °C (range 1.5-5.0 °C). With the reference-less PRF method presented in this thesis using a 3 T scanner and similar voxel size, a mean temperature SD of 0.65 °C was found. An improvement of a factor of 4 over the previously mentioned report can be presented.

A recent study [30] addressed the comparison between the polynomial reference-less MRT and the multi-baseline approach in moving organs, but suffered from a low SNR in the acquired images. The relative deviation between the two methods was found to be of the same range as the standard deviation of the intrinsic noise in the native phase maps. The conclusions from that study however do not apply to the study of this thesis as the implementation of the reference-less method is fundamentally different compared to the polynomial interpolation approach. Moreover, it is considered that intrinsic high SNR of acquired data is a prerequisite for meaningful statistic comparison of results.

Because of the iterative convolution with a fixed kernel, the complexity of the phase reconstruction algorithm is N^3 as a function of the linear size of the domain. A larger domain requires a number of convolutions per step equal to its area, and the number of iterative steps necessary to "penetrate" inside the domain is proportional to the linear size. The method was successfully applied to large ROIs, up to 6.2 cm inner diameter and in all cases the computing time per slice was less than 0.5 s using MATLAB and less than 100 ms using C++ (image calculation environment (ICE, Siemens Healthcare, Erlangen, Germany)).

The Fourier reconstruction of the open border demonstrated stable performance in real situations so long as the border phase information was available along at least 60% of the total perimeter of the domain.

One systematic risk with reference-less PRF MRT is to underestimate the actual temperature inside the domain whenever the border's voxels start to heat up, that is, the border becomes contaminated by the temperature elevation in the tissue. A temperature elevation along the border may occur in the following circumstances: passive heat flow by diffusion (for any heating source), long exposure to a low acoustic field intensity during volumetric HIFU sonication, formation of secondary lobes of the HIFU beam during steering near the border. Such effects are not expected to exceed a few °C, even accumulated, if the border of the domain is chosen carefully. Subsequently, the effectively delivered thermal dose would only moderately exceed the calculated values inside the ROI (i.e. moderate over treatment), without triggering the risk of under-treatment and tumor regrowth. Such a risk was reported whenever a treatment endpoint is to be defined during MR-guided RF ablation based on axial temperature maps [134]. In contrast to the l_1-reweighted polynomial regression [47] the method described here can be applied in such situations where the number of heated voxels is large compared to the total number of voxels contained in the domain used for the polynomial fit. A typical situation is volumetric HIFU sonication [123], or LITT ablation over 10-30 minutes.

Within a clinical scenario, the tumor-to-parenchyma magnetic susceptibility contrast, if significant, may be a source of error for the current implementation of the method, as the fundamental hypothesis here requires a homogeneous or linearly changing distribution of susceptibility. The method as described in this thesis may be adapted to manage for a time-invariant source of magnetic susceptibility, by including a pre-defined term for the local field perturbation as a static correction. The current method does not intrinsically correct for dynamic changes in the bulk susceptibility induced by RF-ablation [134].

Flow compensation is mandatory to avoid misinterpretation of GRE phase shift in large blood vessels as local temperatures change. The sequence used in this thesis for human liver baseline measurements provided in-plane flow compensation. Some residual errors may have occurred from through-plane orientation of large vessels. However, to the level of accuracy of the phase measurement in voxels overlapping large vessels in human liver, it was impossible to distinguish between the residual flow-related phase shift and the intrinsic blood-to-tissue susceptibility contrast. Nevertheless, if defining a different purpose, the same method can be applied to isolate the spatial contrast from the phase map acquired with flow-sensitive sequences (see Figure 4.12 c,d where flow compensation of the acquisition sequence was not available for sheep liver and some large vessels generate positive or negative phase contrast).

The MR sequences and setup, as well as the data processing algorithms for computing PRF temperature maps, are specific to the target organ. In particular, a major difference exists between monitoring moving and non-moving organs. The standard reference-subtraction PRF method works accurately for non-moving organs. For example, an excellent precision of $0.2\,^{\circ}\mathrm{C}$ can be achieved in brain MRT with the head immobilized (see chapter 2). However, the same standard reference approach is confronted with major difficulties in cases where tissue motion occurs between the acquisition time points of the reference and the actual per-operatory image. Any misregistration between the reference and actual images induces an over- or underestimation of the calculated temperature. The worst case is the unique-event (i.e. non-predictable) motion, such as accidental muscle contraction, peristaltic drift of abdominal organs, or swelling/shrinking of tissue induced by the treatment itself. The previously reported multi-baseline approach [28, 29, 135] is not capable of improving the temperature errors induced by non-predictable motion.

The reference-less PRF thermometry published by Rieke et al. [114] is intrinsically insensitive to both periodic and unique-event tissue motion if certain specific border conditions are fulfilled. The previously reported implementations interpolate the spatial information from a rectangular, closed, thick border toward the inner ROI. It may be challenging to place this type of border (additionally required to be unheated and free of artifacts) around the area to be treated in a patient. Moreover, several user interactions are required for that implementation to calculate and display the temperature maps. The l_1-metrics variant published by Grissom et al. [47] relaxes the constraints for the border, but behaves as a spatial high-pass filter and therefore works accurately only for small heated areas under moderate elevation of temperature. Contrarily, in LITT ablations, the heated region is large (several centimeters of size) and the heating regime may frequently exceed $70\,^{\circ}\mathrm{C}$ tissue temperature. Note, neither multi-reference nor reference-less approaches can filter out intra-scan motion artifacts (e.g. ghosting along phase encode direction).

The novel reference-less PRF method developed in this thesis uses a thin, open, and circular border around the heating target area. The calculation of the GRE background phase is based on the solution of the 2D Dirichlet problem. The user has to define the closed or open border once (at the beginning of the intervention), and ulterior redefinition is required only in the case of slice repositioning. No other user interaction is required.

This study presented the first thermo therapy clinical data processed off-line and on-line using the reference-less PRF temperature calculation approach. The novel reference-less method was implemented in the real-time MR reconstruction program (ICE, Siemens Healthcare, Erlangen, Germany) and enabled online monitoring simultaneous to intervention in two conscious patients treated by LITT in liver. The acquisition was performed continuously under free breathing, i.e. no respiratory triggers.

A direct comparison of the near-harmonic reference-less calculation versus the standard subtraction of a reference phase map is also provided as post-processing for respiratory-triggered acquisition in six patients.

Moreover, this is the first clinical study comparing the thermal dose prediction versus posttreatment assessment of ablation when reference-less PRF thermometry is involved. The approach, combining free breathing, respiratory triggered GRE acquisition, near-harmonic reference-less PRF MR thermometry and model-based regularization of temperature values, yielded an estimated precision of $0.7\,°C$ to $2.1\,°C$, resulting in millimeter-range agreement between the calculated thermal dose and the 24h post-treatment unperfused regions in liver in the transverse and sagittal planes.

Note that several recent clinical studies addressed the issue of the thermal dose versus ablative lesion correlation in MR guided RFA. They reported large and systematic overestimation of the true lesion size by the thermal dose-based prediction in the transverse plane [74] with bias mean $11.44 \pm 3.80\,mm$ SD for the maximum diameter and $7.33 \pm 7.14\,mm$ for the minimum diameter, and the absence of any correlation when the imaging plane orientation was varying [132]. These effects are due to thermal cavitation, yielding water steam bubble formation above $100\,°C$ in the vicinity of the RF electrode and producing dynamic changes of tissue susceptibility [134]. For a bipolar electrode, the steam bubbles are distributed around the 1 mm narrow anode-cathode gap [134]. Here, the infrared energy deposition is distributed approximately uniformly along the 30 mm long diffusive tip. The measured temperatures did not exceed the absolute temperature of $90\,°C$, therefore no susceptibility disruptive phenomena were observed.

The assessment of the matching between the ablation region and the initially assigned tumor volume was not defined as an objective of this study, nor correlating the accuracy of the thermal targeting with the patients outcome. The purpose of this work focused on evaluating, under clinical conditions, the capability of correctly measuring the temperature elevation in liver.

Reference-less PRF temperature monitoring enables significant improvement of the temporal resolution (i.e. no respiratory trigger is required). Segmented-EPI GRE acquisition with scanning time of 300 to 350 ms per slice permits the freezing of intra-scan motion while still providing sufficient quality of image (in terms of SNR and low geometric distortion). Therefore the image quality is not traded off against ultra-short temporal resolution [61]. In this study, on-line temperature monitoring could be performed in 3 orthogonal slices under free breathing with 1 s temporal resolution.

Furthermore, reference-less temperature monitoring expands the workflow flexibility. The heat source can be repositioned or additional heat sources could be placed without the time-penalty induced by re-mapping the atlas of tissue motion. The multi-baseline technique is still a time-referenced calculation and requires a waiting period for tissue cooling before starting a new session of temperature monitoring if the slice orientation has changed. Unlike the multi-reference approach, the slice position and orientation may be varied during the intervention when using the reference-less approach.

The reference-less PRF method described in this thesis aims to achieve clinical-standard monitoring of the thermo therapy, but several major steps have to be achieved to enable a clinical routine use. The placement of the border remains difficult in certain cases and requires detailed analysis and profound experience, for example, when the target region is near the liver capsule or if large vessels are nearby. If thermal contamination or artifacts affect the border, the temperature calculation is expected to result in an over- or underestimation. Future work should address the question of using an arbitrary border defined automatically (i.e. user-interaction-free) for an improved practicability and flexibility.

Although the temperature maps calculated by the reference-less method under free breathing and without respiratory trigger were precise at each time point, a spatial registration step is still required to calculate the cumulative thermal dose over the course of the treatment, to be linked to the anatomical frame of coordinates.

In the case of motion of large amplitude and for a fixed ROI (attached to the image coordinate frame), the heated area may reach the ROI's border, violating the unheated border condition. Using an open border may help to reduce this risk. However, additional work is required to register the ROI position with the liver motion.

Several studies in the past discussed the standard reference method and showed its clinical relevance and practicability [17, 18, 66, 74, 87, 109, 111, 112]. The study presented in this thesis demonstrates the potential benefit of the reference-less method (stability, robustness). In particular, in combination with the model-based temporal regularization, the reference-less method provided sufficient stability to enable prediction of tissue necrosis, and sufficient precision to be used as a safety tool to prevent un-prescribed thermal lesions during local thermo therapy.

In conclusion, a novel method for reference-less MR thermometry using the PRF effect is described in this thesis. This is based on a physically meaningful and numerically robust formalism, compatible with real time performance and on-line temperature monitoring. The method was demonstrated to provide accurate baseline measurements in healthy volunteers and accurate calculation of the temperature elevation ex vivo and in moving organs in vivo. The results were nearly identical to the ground-truth measurements ex vivo and in vivo (sheep liver) during HIFU sonication. The offset from the ground-truth data was less than or equal to the standard deviation of the measurement noise. The current limitations are mainly related to the requirement of an artifact-free phase along the border of the domain (or, at least, along sufficient segments of the border), and the requirement for the absence of local sources of bulk susceptibility inside the region of interest.

The reference-less PRF thermometry method based on the theoretical framework of harmonic functions was applied to clinical data acquired during LITT treatments of liver malignancies. This was implemented in the real time MR reconstruction program, and enabled online monitoring during local thermo therapy. The current reference-less MRT approach expands the workflow flexibility, eliminates the need for respiratory triggers, enables higher temporal resolution, and is insensitive to unique-event motion of tissue. Under the condition that no susceptibility-disruptive phenomena occur, this novel reference-less MRT method eliminates the potential temperature errors induced by the swelling/shrinking of heated tissue. Further technical developments are required to enable the routine use of this novel reference-less PRF thermometry method in clinics.

Chapter 5

Outlook: First results using an arbitrary border for reference-less PRF temperature monitoring

The purpose of the current study was the novel implementation of a near-harmonic reference-less PRF thermometry. Chapter 4 describes the boundary value problem for a border, which has to be circular. Within this thesis the new developed near-harmonic reference-less PRF method is extended and enables now the use of a thin and arbitrarily formed border, in particular including nearly the whole liver in the ROI.

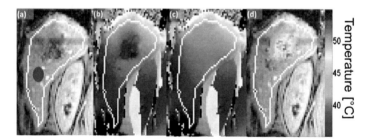

Figure 5.1: *MR images acquired during a treatment of a patients liver using LITT. Image (a) and (d) show magnitude MR images. Image (b) and (c) show unwrapped phase MR images. **(a)** GRE magnitude MR image of the liver. The dark spot located central in the liver is the heated area. The red point placed within this dark spot markers the coordinate used for the graph displayed in image 5.4). The blue data points mark the region used for baseline temperature calculations. The white border placed at the border of the liver defines the voxels, which were taken to get the phase values used for the solution of the boundary value problem to calculate the background phase. **(b)** Shows the GRE native phase map, which provides the phase values on the white border used for the reference-less PRF temperature calculation. **(c)** Shows the reconstructed background phase. **(d)** Shows the calculated temperature $T > 37°C$ overlaid on the magnitude image.*

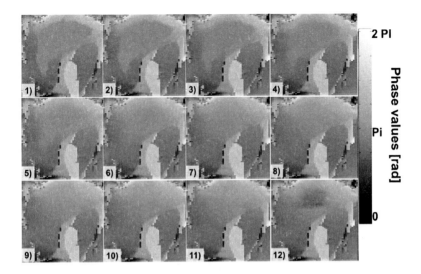

Figure 5.2: *This image visualizes the iterative reconstruction method of the background phase inside the domain of interest. The shown background calculation uses a circular border placed on in vivo liver data acquired during the LITT of a patient. The actual background phase during the reconstruction process is shown before the first iteration in frame 1) (i.e. border information only), and after 1, 5, 10, 20, 40, 100, 200, 500, 700 and 900 iterations in frames 2), 3), 4), 5), 6), 7), 8), 9), 10) and 11) respectively. Whereas, the border data has experimental noise, the reconstructed background is a smooth function. Frame 12) shows the original, unwrapped phase.*

To allow for a reference-less PRF method, which enables the temperature calculation for an arbitrary formed ROI, an improved calculation of the residual value of the 2D phase's Laplacian ϵ has to be provided. Despite the calculation of ϵ, the whole post processing described in section 4.2 can be applied to the improved reference-less temperature calculation method for an arbitrary ROI. To calculate the background (reference) phase within an arbitrary ROI, an ϵ-map has to be provided, not only an average value as described in 4.2.

The ϵ-map was filled using the algorithm described in 4.2.3. For each voxel within the ROI an ϵ-value is calculated using a surrounding circle with a radius of 3 voxels. An example for a resulting ϵ-map is displayed in figure 5.3 3). This ϵ-map includes also non-harmonic parts induced by heating. To exclude the non-harmonic parts, the ϵ-map was fitted linear in two dimensions (Figure 5.3 4)). Using the fitted ϵ-map the background phase and therefore also the relative temperate can be calculated as described in 4.2 (Figure 5.3 4)). The iteratively reconstruction of the background phase using the calculated ϵ-map is exemplary shown in figure 5.2.

A detailed analysis of the improved arbitrary formed border reference-less PRF thermometry was performed on the same patient data used for the evaluation of the reference-less PRF method, which used a circular formed border (see Section 4.3.2 Patient 1-6 in Table 4.2).

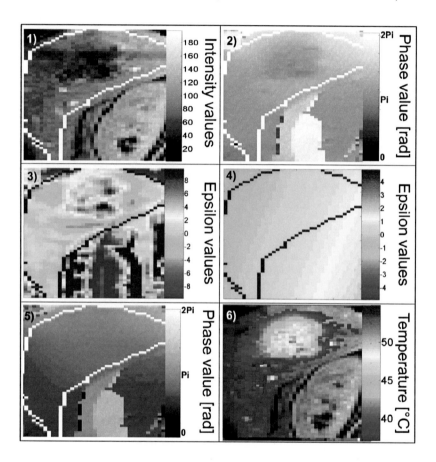

Figure 5.3: *Frame 1) and 2) show the MR magnitude and MR phase image of a liver of a patient during a LITT. The white border on both images marks the ROI for the background (reference) phase calculation. For this ROI an ε-map was calculated, which is displayed in frame 3). The 2D linear fit of the ε-map is shown in frame 4). This fitted ε-map in frame 4) was used calculate the background phase in frame 5) and the relative temperature [°C], which is overlaid on the MR magnitude image in frame 6).*

The temperature elevation in liver was calculated using the standard time-referenced method and also using the reference-less post processing method with a thin border, proximal to the liver margins (Figure 5.1). The closed border circumscribed nearly the whole section of the liver visible in the actual slice. The phase on the border must be smooth, free from artifacts and unheated. The results from the two methods (standard reference and improved reference-less PRF) were compared over the series of patients. The precision of the temperature calculation for each method was assessed using baseline measurements in non-heated ROIs, see Figure 5.1 (a).

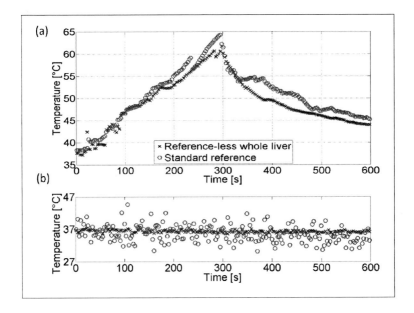

Figure 5.4: *The shown graph displays the temperature development near the laser fiber in one voxel, which is marked as a red dot in image 5.1 (a). Over a period of 300 s the liver tissue is heated. The cooling after treatment was monitored for further 300 s.*
(a) The plotted temperature data are filtered with a multi-exponential pixel-wise model 3.2.3.1 (updated at each sampling point) applied on the measured data using the standard referenced PRF method and, respectively, using the reference-less MRT calculation data. (b) Shows the baseline stability of both methods, which was calculated using the measured data within the blue marked region of image 5.1 (a).

Within the unheated voxels (i.e. baseline calculation; Figure 5.4 (b)), the standard reference post processing method showed an average precision of $\pm 4\,^\circ$C, while the reference-less method greatly improved this precision to $\pm 0.5\,^\circ$C. An example of the temperature vs. time plots in one heated voxel (Figure 5.1 (a), red mark) during the intervention is shown in Figure 5.4 (a). The temperature values are plotted over time for both the standard reference method and the reference-less method. The absolute offset of the standard temperature measurement in the heated regions compared to the more robust reference-less method was $5.6\,^\circ$C averaged over the 6 patients. The main advantages of the reference-less PRF method are: robustness against interscan motion and per-operatory flexibility due to the absence of a reference image.

By using the current reference-less PRF method, the ROI for temperature monitoring can be repositioned/resized and the slice orientation/position can be changed during the intervention. As long as the border of the domain is unheated and comprised inside the liver, the actual implementation does not demand any conditions regarding its size or shape.

The reference-less PRF method using a thin border proximal to the liver margins estimates the temperature elevation with significantly better precision than the standard reference method. In addition to known advantages over interscan motion and external perturbation of the magnetic field, this implementation offers higher per-operatory flexibility for MR thermometry and is compatible with a no-user-interaction scenario for reference-less temperature monitoring over the entire liver if automatic organ segmentation could be provided.

Table 5.1: *Overview of treated lesions.*

ID #	Tumor size (cm) axial	Necrosis size (cm) sagittal	Necrosis size (cm) axial	Necrosis size (cm) sagittal	Liver seg- ment	Nb. of appli- cators	Laser energy (kJ)	Tumor histo- logy
1	3.8x3.3	3.5x2.7	4.3x4.3	4.1x4.0	8	3	48.1	GC
2	2.2x2.1	2.5x2.2	3.3x3.1	3.1x2.9	4	2	32	RC
3	5.0x2.7	4.9x3.3	4.9x4.2	4.8x4.4	8	3	37	GC
4	2.0x1.9	2.2x2.1	4.0x3.5	4.1x3.2	8	2	32.7	CM
5	5.0x3.8	4.8x3.1	4.8x3.1	5.3x3.1	2	3	44	RC
6	5.2x4.8	4.8x3.7	5.0x4.0	5.0x4.1	4	2	27.5	BC

GC=gastric carcinoma, RC=rectum carcinoma, CM=colorectal metastases, BC=breast carcinoma

Chapter 6

Conclusion

The history of MR started with Otto Stern and Walther Gerlach. They showed first that atoms have discrete settings for the magnetic moment. Bloch and Purcell used this knowledge to perform the first successful MR experiments. Since the beginning of MR in the 20th century, an interdisciplinary society grew up and improved MRI year for year. This society encompasses many areas of science and technology, including spin physics, biophysics, image reconstruction, and hardware design. The growing of MRI applications in medical science is amazing, because MRI can be used for diagnostics as well as for interventions in patients.

Ablative treatments are gaining increasing attention as an alternative to standard surgical therapies, especially for patients with contraindications or those who refuse open surgery. Thermal ablation is used in clinical applications mainly for treating heart arrhythmias, benign prostate hyperplasia, and non-operable liver tumors and metastases; there is also increasing use of applications for other organs, including the kidney, lung, and brain. Potential benefits of thermal ablation include reduced morbidity and mortality in comparison with standard surgical resection and the ability to treat nonsurgical patients.

As described by the Fermi-Dirac-Statistic, the temperature correlates with the determination of the average number of fermions in a single-particle state. Therefore, MRI can be used to acquire non-invasively the temperature within the human body during an intervention. The standard temperature MR monitoring method requires the acquisition of two MR images at two different time-points. One image of the native human body to acquire the reference (background) and one image from the heated human body are necessary. Nevertheless, acquiring two images of a moving organ within the human body at two different time points leads to unavoidable errors.

Therefore, a physically and mathematically motivated method is presented to calculate temperature information using a single MR image. A fundamental basis for reference-less MRT is the harmonic behavior of the MR phase within a non-heated homogeneous organ, which means that the Laplace equation can be applied to the harmonic MR phase. Solving the Laplace equation by using the phase on a non-heated border around the treated ROI to calculate the background (reference) phase is a boundary value problem called Dirichlet problem.

Several pre-experiments were necessary for the detailed analysis of the requirements and preconditions have to be fulfilled for a reference-less PRF temperature monitoring method. These pre-studies are discussed in chapter 2 and chapter 3. Chapter 2 handles the standard reference PRF temperature monitoring at non-moving organs and chapter 3 analyzes the standard reference PRF method at moving and deforming organs.

At the human brain, as non-moving organ, three different sequences, the GRE, seg EPI and ss EPI sequence, were discussed in combination with two different field strengths of 1.5 T and 3 T. The evaluations presented in this thesis were performed with a high temporal and spatial resolution. They show a strong dependency on the temperature precision, field strength and echo time and that the echo time should be chosen similar to T_2^* for an optimized precision. Therefore, a compromise between short measuring time, high resolution, and accuracy of the temperature scale (long TE) was necessary. For the ss EPI sequence, the TE was increased until the T_2^* relaxation time was achieved, in order to maximize the SNR and CNR, which results in a precision of 0.16 °C for a TE of 50 ms at 3 T and a precision of 0.25 °C for a TE of 70 ms at 1.5 T. The GRE / seg EPI reach a precision of 0.37 °C / 0.39 °C at 3 T and 0.58 °C / 0.63 °C at 1.5 T with more than twice the acquisition time of the ss EPI sequence. Particularly, the ss EPI provides more than twice the precision of both the GRE and seg EPI sequences in half the acquisition time at both field strengths. On the 3 T system, the precision of all three sequences improves by a factor of 1.5 compared to the 1.5 T system (Figure 2.7, Tables 2.4 and 2.3). Precision, accuracy and corresponding spatial and temporal resolution of the ss EPI enable high quality temperature monitoring simultaneously with HIFU, RFA and LITT of the brain.

Most relevant to this thesis is the temperature monitoring of moving and deforming organs, i.e. in the liver. Temperature monitoring using the standard reference PRF method to monitor the temperature of the liver, as a moving and deforming organ, is much more challenging than on the brain. Data from volunteers as well as from patients has been examined in this thesis. The precision in unheated volunteer livers was in the range of 1 °C to 2 °C; in patients' livers in a non-heated ROI, the precision varies between 1 °C and 8 °C (median: 4.9 ± 4.5 °C). The main reason for the difference in temperature precision between volunteers and patients is that patients, unlike the volunteers, were undergoing a percutaneous intervention, and, despite local anesthesia, had some pain and thus some irregular breathing during the procedure. This irregular breathing can lead to inter- and intra-scan motion artifacts, which typically result in errors of the calculated temperatures. The irregular breathing results in irregular liver displacement and deformation. Therefore, the same slice of the liver with the same deformation can not always be measured in both the reference image and the phase image that contains the information regarding the temperature elevation. This causes either under- or overestimation of temperature. Within this thesis, an analysis was performed including LITT of 34 lesions in 18 patients with simultaneous real-time visualization of relative temperatures. Correlative contrast-enhanced T1-weighted MR images of the liver were acquired 24h after treatment using the same slice positions and angulations as the thermal images acquired during LITT. For each slice, a temperature map and follow-up images were registered for comparison.

Afterward, segmentation was performed based on the temperature $T \geq 52\,^{\circ}C$ at the end of the intervention, and based on necrosis 24h after LITT seen on follow-up images. The error of 13% may have several causes, which can be divided into two different categories, the post processing step and temperature acquisition itself. Errors occurring as a result of the post processing steps are misregistration, missegmentation, and fit $T(t)$ (Equation 4.47). Those resulting from temperature acquisition include: a mismatch between reference and TMap images (e.g. irregular breathing, non-rigid deformation of the liver), non-corrected intrascan motion artifacts (e.g. blood flow, respiratory motion, cardiac motion), and difficulties in the positioning of the ROI for B_0-correction.

The greatest source of error was probably liver motion and irregular deformation of the liver. These factors made slice and voxel matching, as needed for standard reference PRF temperature calculation, challenging.

Table 6.1: *Discussed temperature calculation methods using the PRF method based on MR imaging.*

Standard deviation of residues between end-point fitting curve and directly calculated data		Standard deviation of residues between end-point fitting curve and regularized data	
standard subtraction $[\,^{\circ}C]$	reference-less $[\,^{\circ}C]$	standard subtraction $[\,^{\circ}C]$	reference-less $[\,^{\circ}C]$
10.5±4.2	3.8±1.2	4.5±2.9	1.4±0.6

Temperature monitoring at moving / deforming and non-moving / deforming organs was analyzed. The presented reference-less PRF method showed impressive improvements for the temperature monitoring at the liver as moving and deforming organ.

In this thesis two new methods are presented to reduce motion artifacts. A fit based on the Bioheat equation [16, 104] to describe the temperature distribution in tissue (Section 3.2.3.1) and a reference-less method to monitor temperatures (Chapter 4).

Fitting of phase data takes into account the information of all previously measured time points. Therefore, the fit minimizes artifacts due to outliers in a single measurement and reduces the dispersion of data points. The fit increases certainty and precision of the current data point, because the calculation of the current temperature value includes all temperature values acquired before. Table 4.6 as well as figure 4.13 visualize the benefit of the fit. The precision of the standard reference PRF method was improved from $10.5 \pm 4.2\,^{\circ}C$ to $4.5 \pm 2.9\,^{\circ}C$; those of the reference-less PRF method from $3.8 \pm 1.2\,^{\circ}C$ to $1.4 \pm 0.6\,^{\circ}C$. The improvement by applying the exponential fit showed a more than doubled precision in both cases (Table 6.1). Particularly in combination with the exponential fit (solution of Bioheat equation), both methods (standard reference and reference-less) show a good stability and robustness of temperature calculation, can be used to predict necrosis, and can be used as a safety tool during local thermal therapies.

In the past few years, three different post processing models for PRF methods have been discussed. The standard reference method, which is an approved technique to monitor thermal therapies [112], the multi-baseline approach published first by Sennevall et al. and Vigen et al. [28, 29, 136], and a reference-free method published first by Rieke et al. [114]. Each of these methods has its advantages and disadvantages, which are summarized in table 6.2. Reference-less PRF method cannot be used to monitor temperature during hyperthermia, because the whole body is warmed up to 41 °C to 45 °C. Therefore, for the whole body temperatures have to be monitored during hyperthermia. A reference-free post processing method visualizes temperatures local to one organ. For hyperthermia, the standard reference or the multi baseline approach has to be used.

Rieke et al. published first the idea of a reference-free temperature monitoring. They used a polynomial L_2 interpolation from a rectangular, closed, and thick border, which calculates a reference phase from the border toward the inner ROI [90, 114]. It is hard to place such a rectangular, thick ROI around the treated area in a patient in a way that no heating or artifacts occur at the border. It has to be considered that their mathematical algorithm may lead to numerical singularities such as an ill-conditioned system of equations. Further on, several user interactions are necessary to calculate and display temperatures during thermo therapy, which makes this method incompatible for the clinical practice.

A second method to overcome motion artifacts is the multi-baseline approach [28, 29, 136], which acquires not only one reference phase image, but a set of reference phase images at different positions within the breathing cycle. During intervention, phase images that contain the information regarding the temperature elevation are correlated with the set of reference phase images to find matching images. The multi-baseline approach works in humans, but cannot eliminate interscan motion; it can only reduce them. Artifacts induced by the swelling/shrinking of tissue can not be solved by the multi-baseline approach.

The novel reference-less PRF method developed and evaluated in this thesis, based on the assumption that the measured phase within a homogeneous, not heated tissue, can be described as a harmonic function. This condition allows for solving a boundary problem, the so called Dirichlet problem. The newly developed reference-less method presented in this thesis is not just an interpolation. Contrary to the methods published in the past, this reference-less method based on a mathematical motivated relationship to calculate the background, which is assumed to be a harmonic function of the phase.

The challenge of the reference-less temperature monitoring method is the need of a smooth, harmonic phase without heating or any other artifacts on the border used for the reconstruction of the reference phase around the heated area. Certainty of the reference-less method strongly depends on the chosen border. The smaller and more flexible the border can be, the better it is for the clinical routine. Therefore, the method developed in this thesis uses a thin circular border (1 pixel). Furthermore, the border can be opened once or twice to exclude small heated areas or other artifacts. The open part of the border can be varied in size and position (Figure 6.1).

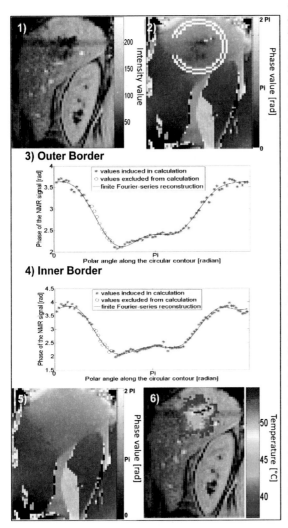

Figure 6.1: *This figure gives a brief impression of the new reference-less PRF method to monitor temperatures on moving and deforming human organs. Part 1) of the figure shows one magnitude MR image of the treated liver. The dark area within the liver describes the heated zone. Part 2) of the figure shows the phase MR image of the same ROI of the liver in the patient. The phase data on the white marked border are used to calculate the background phase based on the harmonic behavior of the MR-phase and on a 2D Dirichlet problem. The white marked border is divided in an outer and an inner part to calculate the average residual value of the 2D phase's Laplacian. Both border are opened to exclude non-harmonic phase's. Therefor, the missing phase data points have to be calculated. Part 3) and 4) of the figure show the phase on the inner and outer border as well as the fitted Fourier series of 1D trigonometric functions, which is used to fill the missing phase data points on the circular border. Part 5) of the figure shows the reconstructed background (reference) phase, which is used to calculate the relative temperature [°C] shown in part 6) of the figure as overlay on the MR magnitude image.*

Table 6.2: *Overview on post processing methods using the PRF method and GRE based sequences at MRI to monitor temperatures*

Case	post processing method	Advantage	Disadvantage
local thermo therapy on non-moving organs(brain) and phantoms	standard reference PRF method	easy to use; temperature monitoring possible for whole slice	moving or deformation induces over- or underestimation of temperature; reference image needed (less flexibility; no setup changing during treatment possible); B_0-correction needed
local thermo therapy on moving organs (liver, prostate)	reference-less PRF-method	eliminate interscan motion artifacts; can be used in free breathing ⇒ fast; high flexibility (variation of slice orientation and position, of applicator number and position, ROI); no B_0-correction necessary;	temperature monitoring only in ROI; dependency of tempera-pure accuracy and precision from border; need of registration tool for free breathing acquisition for thermal dose; difficult if target region is near the organ border;
	multi-baseline PRF-method	reduces interscan motion artifacts; thermal dose calculation; can be used in free breathing ⇒ fast;	couldn't eliminate interscan motion artifacts; less flexible than reference-less method; B_0-correction necessary;
Hyperthermia	standard reference PRF method	temperature monitoring possible for whole slice	sensitive to motion; for moving and deforming organs triggering or gating necessary
	multi-baseline PRF-method	discussed above	discussed above

The standard reference PRF method within the human body in moving as well as non-moving organs shows a temperature precision less than the theoretical possible temperature precision. For non-moving organs, a high SNR and therefore a high theoretical temperature precision can be achieved, which was determined to be 0.04 °C for a SNR of 65. For moving organs, a higher temporal resolution for the MRI acquisition is required and therefore the achievable SNR is lower. An SNR of 17 in the liver, as a moving and deforming organ, corresponds to the theoretical temperature precision of 1.35 °C. The standard reference PRF method achieved a temperature precision of 4.5±2.4 °C. In comparison, the new reference-less PRF method presented in this thesis shows a temperature precision of 1.45±0.3 °C, which is in the same range than the theoretical possible value of 1.35 °C (Table 6.3).

Table 6.3: *Improvement obtained by the exponential fit discussed in 3.2.3.1.*

PRF method	organ	SNR	STD [°C]	temperature precision [°C]
standard reference	brain	65± 7	0.041± 0.003	0.16± 0.06
standard reference	liver	17± 2	1.347± 0.201	4.48± 2.36
reference-less	liver	17± 2	1.347± 0.201	1.45±0.28

The exponential fit is a solution of the Bioheat equation proposed by Pennes. Patient data were acquired during LITT in the liver and post processed twice. Once to calculate the temperature using the standard reference PRF method and once using the reference-less PRF method. Both results, from the standard temperature monitoring and the new developed one, were corrected using the exponential fit. Corrected temperature values were plotted over time voxel by voxel. Additional, the exponential fit at the end of the treatment was calculated using the information of the whole duration of the intervention (end-point fitting). The standard deviation between corrected (regularized) / directly calculated and the end-point fitting curve was calculated. Further on, the theoretical SD (STD), assuming that white noise is the only source of error, was calculated.

The non-invasive MRT is used as a safety tool for local thermal therapies (LITT, RFA, HIFU) to predict the necrosis. At the same time, healthy tissue around the treated organ has to be protected and all malignant tissue including a safety border has to be destroyed. The tissue state can be estimated using the peak temperature or the thermal dose. Especially the thermal dose calculation needs a high temperature precision. Even small deviations of temperature can lead to over- or underestimations of the induced necrosis. In particular, HIFU induces a temperature rise of several degrees in a few seconds and requires fast and precise temperature monitoring with a frequency less than 1 s, which can be provided using the novel reference-less PRF method presented in this thesis for non-moving as well as for moving and deforming organs.

The novel reference-less PRF method developed in this thesis is a promising tool to calculate temperatures in real-time on non-moving organs, enabling higher flexibility for the user. This novel reference-less method enables real-time monitoring simultaneously with a thermo therapy. During an intervention only one user-interaction is needed. An open border has to be placed and defined one time at the beginning of the intervention (Figure 6.1 2)). No other user interaction is needed. Calculated temperature values are displayed as overlay in an online window during intervention on the magnitude image (Figure 6.1 6)). Even so, further steps have to be done to enable easier operation. The placement of the border is, in some cases, difficult in patients, for example, if the target region is near the liver capsule. An arbitrary border enables a better practicability and further flexibility.

This novel reference-less PRF method allows for temperature monitoring without any breathing trigger. The whole breathing cycle can be used for temperature monitoring, which enables real-time temperature monitoring with a temporal resolution of 300 ms. One disadvantage is that an acquisition without any breathing trigger requires a good registration tool to calculate a thermal dose. Future work should include an implementation of a registration tool. Free breathing also leads to problems when the border is restricted in size by surrounding artifacts (vessel, liver capsule). The border is fixed on a certain coordinate within the image slice, but the liver moves through the image slice in free breathing. The artifacts or heated area can move into the border. In future work, a liver segmentation method will be shown, which allows to shift the ROI with the liver movement.

In summary, within this thesis an analysis of the standard reference PRF method on non-moving as well as on moving and deforming organs was performed. Two improvements are presented. A voxel-by-voxel exponential fit of temperature data over time and a reference-less PRF method. Both post processing methods work fine and can be combined. The fit regularizes the measured data for an easier interpretation and a better reliability for clinical exercise, while the reference-less PRF method solves and eliminates the main reasons for temperature over or under estimation. No reference-image and therefore also no B_0-correction is needed, if temperature data is acquired with the reference-less method. Temperature maps are calculated using just one phase image, which is acquired during the intervention. Therefore, the reference-less PRF method eliminates the interscan motion artifacts, these are usually the main cause for temperature errors using the standard reference PRF method. Furthermore, the reference-less PRF method allows for more flexibility than the standard reference method. Slice orientation / angulations can be varied during the intervention. Applicators can be repositioned or additional applicators can be positioned during the intervention. One great benefit of the reference-less method is the fast acquisition. No breathing triggering is needed and temperature monitoring can be done in free breathing.

Bibliography

[1] A. Abragam. *The principles of nuclear magnetism*. The Claredon Press, 1983. 20

[2] L. J. Anderson, S. Holden, B. Davis, E. Prescott, C. C. Charrier, N. H. Bunce, D. N. Firmin, B. Wonke, J. Porter, J. M. Walker, and D. J. Pennell. Cardiovascular T2-star (T2*) magnetic resonance for the early diagnosis of myocardial iron overload. *Eur. Heart J.*, 22:2171–2179, Dec 2001. 57

[3] S. Arrhenius. Ueber die Reaktionsgeschwindigkeit bei der Inversion von Rohrzucker durch Saeuren. *Zeitsch. physikal. Chem.*, 22:221–252, 1889. 31

[4] S. A. Bernard, T. W. Gray, M. D. Buist, B. M. Jones, W. Silvester, G. Gutteridge, and K. Smith. Treatment of comatose survivors of out-of-hospital cardiac arrest with induced hypothermia. *N. Engl. J. Med.*, 346:557–563, Feb 2002. 35

[5] J. M. Bland and D. G. Altman. Statistical methods for assessing agreement between two methods of clinical measurement. *Lancet*, 1:307–310, Feb 1986. 85, 95

[6] A. R. Bleier, F. A. Jolesz, M. S. Cohen, R. M. Weisskoff, J. J. Dalcanton, N. Higuchi, D. A. Feinberg, B. R. Rosen, R. C. McKinstry, and S. G. Hushek. Real-time magnetic resonance imaging of laser heat deposition in tissue. *Magn Reson Med*, 21:132–137, Sep 1991. 21

[7] N. Bloembergen, E. Purcell, and R. Pound. Relaxation effects in nuclear magnetic resonance absorption. *Physical Review*, 73:679–712, 1948. 22

[8] A. Boss, H. Graf, B. Muller-Bierl, S. Clasen, D. Schmidt, P. L. Pereira, and F. Schick. Magnetic susceptibility effects on the accuracy of MR temperature monitoring by the proton resonance frequency method. *J Magn Reson Imaging*, 22:813–820, Dec 2005. 15, 35

[9] R. M. Botnar, P. Steiner, B. Dubno, P. Erhart, G. K. von Schulthess, and J. F. Debatin. Temperature quantification using the proton frequency shift technique: In vitro and in vivo validation in an open 0.5 tesla interventional MR scanner during RF ablation. *J Magn Reson Imaging*, 13:437–444, Mar 2001. 15, 35

[10] P. A. Bottomley, T. H. Foster, R. E. Argersinger, and L. M. Pfeifer. A review of normal tissue hydrogen NMR relaxation times and relaxation mechanisms from 1-100 MHz:

dependence on tissue type, NMR frequency, temperature, species, excision, and age. *Med Phys*, 11:425–448, 1984. 23

[11] A. Cernicanu, M. Lepetit-Coiffe, J. Roland, C. D. Becker, and S. Terraz. Validation of fast MR thermometry at 1.5 T with gradient-echo echo planar imaging sequences: phantom and clinical feasibility studies. *NMR Biomed*, 21:849–858, Oct 2008. 36, 49, 51

[12] A. Cernicanu, M. Lepetit-Coiffe, J. Roland, C. D. Becker, and S. Terraz. Validation of fast MR thermometry at 1.5 T with gradient-echo echo planar imaging sequences: phantom and clinical feasibility studies. *NMR Biomed*, 21:849–858, Oct 2008. 56

[13] J. Chen, B. L. Daniel, and K. B. Pauly. Investigation of proton density for measuring tissue temperature. *J Magn Reson Imaging*, 23:430–434, Mar 2006. 20

[14] L. Chen, J. P. Wansapura, G. Heit, and K. Butts. Study of laser ablation in the in vivo rabbit brain with MR thermometry. *J Magn Reson Imaging*, 16:147–152, Aug 2002. x, 61, 69

[15] H. L. Cheng and D. B. Plewes. Tissue thermal conductivity by magnetic resonance thermometry and focused ultrasound heating. *J Magn Reson Imaging*, 16:598–609, Nov 2002. 29

[16] H. L. Cheng and D. B. Plewes. Tissue thermal conductivity by magnetic resonance thermometry and focused ultrasound heating. *J Magn Reson Imaging*, 16:598–609, Nov 2002. 59, 94, 123

[17] R. Chopra, K. Tang, M. Burtnyk, A. Boyes, L. Sugar, S. Appu, L. Klotz, and M. Bronskill. Analysis of the spatial and temporal accuracy of heating in the prostate gland using transurethral ultrasound therapy and active MR temperature feedback. *Phys Med Biol*, 54:2615–2633, May 2009. 73, 114

[18] S. Clasen and P. L. Pereira. Magnetic resonance guidance for radiofrequency ablation of liver tumors. *J Magn Reson Imaging*, 27:421–433, Feb 2008. x, 15, 56, 73, 114

[19] H. E. Cline, K. Hynynen, C. J. Hardy, R. D. Watkins, J. F. Schenck, and F. A. Jolesz. MR temperature mapping of focused ultrasound surgery. *Magn Reson Med*, 31:628–636, Jun 1994. 22

[20] H. E. Cline, K. Hynynen, E. Schneider, C. J. Hardy, S. E. Maier, R. D. Watkins, and F. A. Jolesz. Simultaneous magnetic resonance phase and magnitude temperature maps in muscle. *Magn Reson Med*, 35:309–315, Mar 1996. 27

[21] H. E. Cline, J. F. Schenck, R. D. Watkins, K. Hynynen, and F. A. Jolesz. Magnetic resonance-guided thermal surgery. *Magn Reson Med*, 30:98–106, Jul 1993. 22

[22] T. E. Conturo and G. D. Smith. Signal-to-noise in phase angle reconstruction: dynamic range extension using phase reference offsets. *Magn Reson Med*, 15:420–437, Sep 1990. 93, 94

[23] J. De Poorter. Noninvasive MRI thermometry with the proton resonance frequency method: study of susceptibility effects. *Magn Reson Med*, 34:359–367, Sep 1995. 26

[24] J. De Poorter, C. De Wagter, Y. De Deene, C. Thomsen, F. Stahlberg, and E. Achten. Noninvasive MRI thermometry with the proton resonance frequency (PRF) method: in vivo results in human muscle. *Magn Reson Med*, 33:74–81, Jan 1995. 27

[25] J. De Poorter, C. De Wagter, Y. De Deene, C. Thomsen, F. Stahlberg, and E. Achten. Noninvasive MRI thermometry with the proton resonance frequency (PRF) method: in vivo results in human muscle. *Magn Reson Med*, 33:74–81, Jan 1995. 27

[26] J. De Poorter, C. De Wagter, Y. De Deene, C. Thomsen, F. Stahlberg, and E. Achten. Noninvasive MRI thermometry with the proton resonance frequency (PRF) method: in vivo results in human muscle. *Magn Reson Med*, 33:74–81, Jan 1995. 38

[27] B. D. de Senneville, C. Mougenot, and C. T. Moonen. Real-time adaptive methods for treatment of mobile organs by MRI-controlled high-intensity focused ultrasound. *Magn Reson Med*, 57:319–330, Feb 2007. 70, 73

[28] B. D. de Senneville, C. Mougenot, and C. T. Moonen. Real-time adaptive methods for treatment of mobile organs by MRI-controlled high-intensity focused ultrasound. *Magn Reson Med*, 57:319–330, Feb 2007. 112, 124

[29] B. D. de Senneville, C. Mougenot, B. Quesson, I. Dragonu, N. Grenier, and C. T. Moonen. MR thermometry for monitoring tumor ablation. *Eur Radiol*, 17:2401–2410, Sep 2007. 73, 112, 124

[30] B. D. de Senneville, S. Roujol, C. Moonen, and M. Ries. Motion correction in MR thermometry of abdominal organs: a comparison of the referenceless vs. the multibaseline approach. *Magn Reson Med*, 64:1373–1381, Nov 2010. 110

[31] J. A. de Zwart, P. van Gelderen, D. J. Kelly, and C. T. Moonen. Fast magnetic-resonance temperature imaging. *J Magn Reson B*, 112:86–90, Jul 1996. 38

[32] C. J. Diederich, W. H. Nau, A. B. Ross, P. D. Tyreus, K. Butts, V. Rieke, and G. Sommer. Catheter-based ultrasound applicators for selective thermal ablation: progress towards MRI-guided applications in prostate. *Int J Hyperthermia*, 20:739–756, Nov 2004. 15, 35

[33] O. Dietrich, J. G. Raya, S. B. Reeder, M. F. Reiser, and S. O. Schoenberg. Measurement of signal-to-noise ratios in MR images: influence of multichannel coils, parallel imaging, and reconstruction filters. *J Magn Reson Imaging*, 26:375–385, Aug 2007. 41

[34] O. Dietrich, J. G. Raya, S. B. Reeder, M. F. Reiser, and S. O. Schoenberg. Measurement of signal-to-noise ratios in MR images: influence of multichannel coils, parallel imaging, and reconstruction filters. *J Magn Reson Imaging*, 26:375–385, Aug 2007. 93

[35] P. Dirac. The Quantum Theory of the Electron. *Proc. Roy. Soc. London*, A117:610–624, 1928. x

[36] P. Dirac. The Quantum Theory of the Electron, Part II. *Proc. Roy. Soc. London*, A118:351–361, 1928. x

[37] P. Dirac. *The Principle of Quantum Mechanics*. Oxford: Clarendon Press, 1930. x

[38] A. M. El-Sharkawy, M. Schar, P. A. Bottomley, and E. Atalar. Monitoring and correcting spatio-temporal variations of the MR scanner's static magnetic field. *MAGMA*, 19:223–236, Nov 2006. 27

[39] G. K. Eyrich, E. Bruder, P. Hilfiker, B. Dubno, H. H. Quick, M. A. Patak, K. W. Gratz, and H. F. Sailer. Temperature mapping of magnetic resonance-guided laser interstitial thermal therapy (LITT) in lymphangiomas of the head and neck. *Lasers Surg Med*, 26:467–476, 2000. x, 35, 36

[40] M. J. Firbank, A. Coulthard, R. M. Harrison, and E. D. Williams. A comparison of two methods for measuring the signal to noise ratio on MR images. *Phys Med Biol*, 44:N261–264, Dec 1999. 38

[41] M. J. Firbank, A. Coulthard, R. M. Harrison, and E. D. Williams. A comparison of two methods for measuring the signal to noise ratio on MR images. *Phys Med Biol*, 44:N261–264, Dec 1999. 41

[42] M. J. Firbank, A. Coulthard, R. M. Harrison, and E. D. Williams. A comparison of two methods for measuring the signal to noise ratio on MR images. *Phys Med Biol*, 44:N261–264, Dec 1999. 93

[43] J. Frahm, A. Haase, and D. Matthaei. Rapid NMR imaging of dynamic processes using the FLASH technique. *Magn Reson Med*, 3:321–327, Apr 1986. 9

[44] C. Gabriel, A. Peyman, and E. H. Grant. Electrical conductivity of tissue at frequencies below 1 MHz. *Phys Med Biol*, 54:4863–4878, Aug 2009. 84

[45] W. Gerlach and O. Stern. Der experimentelle Nachweis der Richtungsquantelung im Magnetfeld. *Zeitschrift fuer Physik A Hadrons and Nuclei*, 9:349–352. x

[46] W. Gerlach and O. Stern. Das magnetische Moment des Silberatoms. *Zeitschrift fuer Physik A Hadrons and Nuclei*, 9:353–355, 1922. 10.1007/BF01326984. x

[47] W. A. Grissom, M. Lustig, A. B. Holbrook, V. Rieke, J. M. Pauly, and K. Butts-Pauly. Reweighted L1 referenceless PRF shift thermometry. *Magn Reson Med*, 64:1068–1077, Oct 2010. 74, 111, 112

[48] D. H. Gultekin and J. C. Gore. Temperature dependence of nuclear magnetization and relaxation. *J. Magn. Reson.*, 172:133–141, Jan 2005. 20

[49] E. Haacke, R. Brown, M. Thompson, and R. Ventatesan. *Magnetic Resonance Imaging: Physical Principles and Sequence Design.* Willey-Liss, 1999. 5

[50] R. W. Habash, R. Bansal, D. Krewski, and H. T. Alhafid. Thermal therapy, part 1: an introduction to thermal therapy. *Crit Rev Biomed Eng*, 34:459–489, 2006. 33

[51] R. W. Habash, R. Bansal, D. Krewski, and H. T. Alhafid. Thermal therapy, Part III: ablation techniques. *Crit Rev Biomed Eng*, 35:37–121, 2007. 16

[52] E. L. Hahn. Spin echos. *Phys Rev*, 80:580–594, 1950. 9

[53] J. Hindman. Proton resonance shift of water in gas and liquid states. *Journal of Chemical Physics*, 44:4582–4592, 1966. 24, 36, 73

[54] S. L. Hokland, M. Pedersen, R. Salomir, B. Quesson, H. St?dkilde-J?rgensen, and C. T. Moonen. MRI-guided focused ultrasound: methodology and applications. *IEEE Trans Med Imaging*, 25:723–731, Jun 2006. 15, 35

[55] P. E. Huppert, J. Trubenbach, F. Schick, P. Pereira, C. Konig, and C. D. Claussen. [MRI-guided percutaneous radiofrequency ablation of hepatic neoplasms-first technical and clinical experiences]. *RöFo*, 172:692–700, Aug 2000. x, 15

[56] K. Hynynen, G. T. Clement, N. McDannold, N. Vykhodtseva, R. King, P. J. White, S. Vitek, and F. A. Jolesz. 500-element ultrasound phased array system for noninvasive focal surgery of the brain: a preliminary rabbit study with ex vivo human skulls. *Magn Reson Med*, 52:100–107, Jul 2004. x, 36

[57] K. Hynynen, N. McDannold, R. V. Mulkern, and F. A. Jolesz. Temperature monitoring in fat with MRI. *Magn Reson Med*, 43:901–904, Jun 2000. 22

[58] Y. Ishihara, A. Calderon, H. Watanabe, K. Okamoto, Y. Suzuki, K. Kuroda, and Y. Suzuki. A precise and fast temperature mapping using water proton chemical shift. *Magn Reson Med*, 34:814–823, Dec 1995. 24, 26, 36, 73

[59] F. Johnson, H. Eyring, and B. Stover. *Theory of rate processes in biology and medicine.* John Wiley, 1974. 20

[60] F. A. Jolesz and N. McDannold. Current status and future potential of MRI-guided focused ultrasound surgery. *J Magn Reson Imaging*, 27:391–399, Feb 2008. x, 15

[61] A. Kickhefel, J. Roland, C. Weiss, and F. Schick. Accuracy of real-time MR temperature mapping in the brain: a comparison of fast sequences. *Phys Med*, 26:192–201, Oct 2010. 113

[62] J. H. Kim and E. W. Hahn. Clinical and biological studies of localized hyperthermia. *Cancer Res.*, 39:2258–2261, Jun 1979. 35

[63] C. Kittel and H. Kroemer. *Thermal Physics.* San Francisco: W. H. Freeman, 1980. x

[64] M. S. Kotys, E. Akbudak, J. Markham, and T. E. Conturo. Precision, signal-to-noise ratio, and dose optimization of magnitude and phase arterial input functions in dynamic susceptibility contrast MRI. *J Magn Reson Imaging*, 25:598–611, Mar 2007. 36, 38

[65] D.-J. Kroon. 61

[66] K. Kuroda. Non-invasive MR thermography using the water proton chemical shift. *Int J Hyperthermia*, 21:547–560, Sep 2005. 21, 73, 114

[67] K. Kuroda. Non-invasive MR thermography using the water proton chemical shift. *Int J Hyperthermia*, 21:547–560, Sep 2005. 36

[68] K. Kuroda, D. Kokuryo, E. Kumamoto, K. Suzuki, Y. Matsuoka, and B. Keserci. Optimization of self-reference thermometry using complex field estimation. *Magn Reson Med*, 56:835–843, Oct 2006. 70

[69] K. Kuroda, R. V. Mulkern, K. Oshio, L. P. Panych, T. Nakai, T. Moriya, S. Okuda, K. Hynynen, and F. A. Jolesz. Temperature mapping using the water proton chemical shift: self-referenced method with echo-planar spectroscopic imaging. *Magn Reson Med*, 44:167, Jul 2000. 21, 27

[70] K. Kuroda, R. V. Mulkern, K. Oshio, L. P. Panych, T. Nakai, T. Moriya, S. Okuda, K. Hynynen, F. A. Jolesz, and F. A. Joles. Temperature mapping using the water proton chemical shift: self-referenced method with echo-planar spectroscopic imaging. *Magn Reson Med*, 43:220–225, Feb 2000. 38

[71] K. Kuroda, K. Oshio, A. H. Chung, K. Hynynen, and F. A. Jolesz. Temperature mapping using the water proton chemical shift: a chemical shift selective phase mapping method. *Magn Reson Med*, 38:845–851, Nov 1997. 21

[72] P. Lauterbur, D. N. Levin, and R. B. Marr. Theory and Simulation of NMR Spectroscopic Imaging and Field Plotting by Projection Reconstruction Involving an Intrinsic Frequency Dimension. *J. Magn. Reson.*, 59:536–541, 1984. x

[73] D. Le Bihan, J. Delannoy, and R. L. Levin. Temperature mapping with MR imaging of molecular diffusion: application to hyperthermia. *Radiology*, 171:853–857, Jun 1989. 21

[74] M. Lepetit-Coiffe, H. Laumonier, O. Seror, B. Quesson, M. B. Sesay, C. T. Moonen, N. Grenier, and H. Trillaud. Real-time monitoring of radiofrequency ablation of liver tumors using thermal-dose calculation by MR temperature imaging: initial results in nine patients, including follow-up. *Eur Radiol*, 20:193–201, Jan 2010. 113, 114

[75] C. J. Lewa and Z. Majewska. Temperature relationships of proton spin-lattice relaxation time T1 in biological tissues. *Bull Cancer*, 67:525–530, 1980. 22

[76] L. Lin. A concordance correlation coefficiant to evaluate reproducibilityeasibility. *Biometrics*, 45:255–268, 1989. 85

[77] L. Lin. A note on the concordance correlation coefficiant. *Biometrics*, 56:324–325, Jan 2000. 85

[78] R. N. Low, G. D. Alzate, and A. Shimakawa. Motion suppression in MR imaging of the liver: comparison of respiratory-triggered and nontriggered fast spin-echo sequences. *AJR Am J Roentgenol*, 168:225–231, Jan 1997. 73

[79] M. G. Mack, R. Straub, K. Eichler, K. Engelmann, S. Zangos, A. Roggan, D. Woitaschek, M. Bottger, and T. J. Vogl. Percutaneous MR imaging-guided laser-induced thermotherapy of hepatic metastases. *Abdom Imaging*, 26:369–374, 2001. x, 15, 35

[80] M. G. Mack and T. J. Vogl. MR-guided ablation of head and neck tumors. *Magn Reson Imaging Clin N Am*, 10:707–713, Nov 2002. x, 15

[81] M. G. Mack and T. J. Vogl. MR-guided ablation of head and neck tumors. *Magn Reson Imaging Clin N Am*, 10:707–713, Nov 2002. 35, 36

[82] P. Mansfield. Multi-planar image formation using NMR spin echoes. *J Phys C: Solid State Phys*, 10:L55–58, 1977. 11

[83] P. Mansfield, R. Bowtell, S. Blackband, and D. N. Guilfoyle. Magnetic resonance imaging: applications of novel methods in studies of porous media. *Magn Reson Imaging*, 10:741–746, 1992. x

[84] P. Mansfield, P. R. Harvey, and S. M. K. Echo-volumar imaging. *MAGMA*, 2:291–294, 1994. x

[85] I. Marshall, B. Karaszewski, J. M. Wardlaw, V. Cvoro, K. Wartolowska, P. A. Armitage, T. Carpenter, M. E. Bastin, A. Farrall, and K. Haga. Measurement of regional brain temperature using proton spectroscopic imaging: validation and application to acute ischemic stroke. *Magn Reson Imaging*, 24:699–706, Jul 2006. 21

[86] R. Matsumoto, K. Oshio, and F. A. Jolesz. Monitoring of laser and freezing-induced ablation in the liver with T1-weighted MR imaging. *J Magn Reson Imaging*, 2:555–562, 1992. 22

[87] N. McDannold. Quantitative MRI-based temperature mapping based on the proton resonant frequency shift: review of validation studies. *Int J Hyperthermia*, 21:533–546, Sep 2005. 36, 73, 114

[88] N. McDannold, G. Clement, E. Zadicario, P. Black, F. Jolesz, and K. Hynynen. Transcarnial MRI-guided focused surgery of brain tumours: Initial findings in patients. *Proc Intl Soc Mag Reson Med*. 36, 37

[89] N. McDannold, K. Hynynen, and F. Jolesz. MRI monitoring of the thermal ablation of tissue: effects of long exposure times. *J Magn Reson Imaging*, 13:421–427, Mar 2001. 28

[90] N. McDannold, C. Tempany, F. Jolesz, and K. Hynynen. Evaluation of referenceless thermometry in MRI-guided focused ultrasound surgery of uterine fibroids. *J Magn Reson Imaging*, 28:1026–1032, Oct 2008. 124

[91] D. Meister, F. Hubner, M. Mack, and T. J. Vogl. MR thermometry for laser-induced thermotherapy at 1.5 Tesla. *RöFo*, 179:497–505, May 2007. 51, 52

[92] J. A. Moriarty, J. C. Chen, C. M. Purcell, L. C. Ang, R. S. Hinks, R. D. Peters, R. M. Henkelman, D. B. Plewes, M. J. Bronskill, and W. Kucharczyk. MRI monitoring of interstitial microwave-induced heating and thermal lesions in rabbit brain in vivo. *J Magn Reson Imaging*, 8:128–135, 1998. 15

[93] D. Morvan, A. Leroy-Willig, A. Malgouyres, C. A. Cuenod, P. Jehenson, and A. Syrota. Simultaneous temperature and regional blood volume measurements in human muscle using an MRI fast diffusion technique. *Magn Reson Med*, 29:371–377, Mar 1993. 21

[94] M. E. Moseley, Y. Cohen, J. Mintorovitch, L. Chileuitt, H. Shimizu, J. Kucharczyk, M. F. Wendland, and P. R. Weinstein. Early detection of regional cerebral ischemia in cats: comparison of diffusion- and T2-weighted MRI and spectroscopy. *Magn Reson Med*, 14:330–346, May 1990. 21

[95] C. Mougenot, B. Quesson, B. D. de Senneville, P. L. de Oliveira, S. Sprinkhuizen, J. Palussiere, N. Grenier, and C. T. Moonen. Three-dimensional spatial and temporal temperature control with MR thermometry-guided focused ultrasound (MRgHIFU). *Magn Reson Med*, 61:603–614, Mar 2009. x, 36, 37

[96] C. Mougenot, B. Quesson, B. D. de Senneville, P. L. de Oliveira, S. Sprinkhuizen, J. Palussiere, N. Grenier, and C. T. Moonen. Three-dimensional spatial and temporal temperature control with MR thermometry-guided focused ultrasound (MRgHIFU). *Magn Reson Med*, 61:603–614, Mar 2009. 15

[97] U. G. Mueller-Lisse, M. Frimberger, P. Schneede, A. F. Heuck, R. Muschter, and M. F. Reiser. Perioperative prediction by MRI of prostate volume six to twelve months after laser-induced thermotherapy of benign prostatic hyperplasia. *J Magn Reson Imaging*, 13:64–68, Jan 2001. x, 15

[98] M. Murakami, T. Tsukahara, H. Ishikura, T. Hatano, T. Nakakuki, E. Ogino, and T. Aoyama. Successful use of prolonged mild hypothermia in a patient with severe head injury and diffuse brain swelling. Case report. *Neurol. Med. Chir. (Tokyo)*, 47:116–120, Mar 2007. 35

[99] T. R. Nelson and S. M. Tung. Temperature dependence of proton relaxation times in vitro. *Magn Reson Imaging*, 5:189–199, 1987. 23

[100] A. M. Oros-Peusquens, M. Laurila, and N. J. Shah. Magnetic field dependence of the distribution of NMR relaxation times in the living human brain. *Magn Res Mat Physics*, 21:131–147, Mar 2008. 38

[101] J. Palussiere, R. Salomir, B. Le Bail, R. Fawaz, B. Quesson, N. Grenier, and C. T. Moonen. Feasibility of MR-guided focused ultrasound with real-time temperature mapping and continuous sonication for ablation of VX2 carcinoma in rabbit thigh. *Magn Reson Med*, 49:89–98, Jan 2003. 80

[102] D. L. Parker. Applications of NMR imaging in hyperthermia: an evaluation of the potential for localized tissue heating and noninvasive temperature monitoring. *IEEE Trans Biomed Eng*, 31:161–167, Jan 1984. 22

[103] D. L. Parker, V. Smith, P. Sheldon, L. E. Crooks, and L. Fussell. Temperature distribution measurements in two-dimensional NMR imaging. *Med Phys*, 10:321–325, 1983. 22

[104] H. Pennes. Analysis of tissue and arterial blood temperatures in the resting human forearm. *J Appl Physiol*, 1:93–122, 1948. 29, 59, 94, 123

[105] R. D. Peters, E. Chan, J. Trachtenberg, S. Jothy, L. Kapusta, W. Kucharczyk, and R. M. Henkelman. Magnetic resonance thermometry for predicting thermal damage: an application of interstitial laser coagulation in an in vivo canine prostate model. *Magn Reson Med*, 44:873–883, Dec 2000. 61, 69

[106] R. D. Peters, R. S. Hinks, and R. M. Henkelman. Ex vivo tissue-type independence in proton-resonance frequency shift MR thermometry. *Magn Reson Med*, 40:454–459, Sep 1998. 92

[107] R. Puls, C. Stroszczynski, C. Rosenberg, J. P. Kuehn, K. Hegenscheid, U. Speck, A. Stier, and N. Hosten. Three-dimensional gradient-echo imaging for percutaneous MR-guided laser therapy of liver metastasis. *J Magn Reson Imaging*, 25:1174–1178, Jun 2007. x, 15

[108] E. M. Purcell, H. C. Torrey, and R. V. Pound. Resonance Absorption by Nuclear Magnetic Moments in a Solid. *Phys. Rev.*, 69(1-2):37–38, Jan 1946. x

[109] B. Quesson, J. A. de Zwart, and C. T. Moonen. Magnetic resonance temperature imaging for guidance of thermotherapy. *J Magn Reson Imaging*, 12:525–533, Oct 2000. x, 36, 39, 52, 57, 73, 114

[110] H. Rempp, S. Clasen, A. Boss, J. Roland, A. Kickhefel, C. Schraml, C. D. Claussen, F. Schick, and P. L. Pereira. Prediction of cell necrosis with sequential temperature mapping after radiofrequency ablation. *J Magn Reson Imaging*, 30:631–639, Sep 2009. 70

[111] H. Rempp, P. Martirosian, A. Boss, S. Clasen, A. Kickhefel, M. Kraiger, C. Schraml, C. Claussen, P. Pereira, and F. Schick. MR temperature monitoring applying the proton resonance frequency method in liver and kidney at 0.2 and 1.5 T: segment-specific attainable precision and breathing influence. *Magn Res Mat Physics*, 21:333–343, Sep 2008. 15, 56, 73, 114

[112] V. Rieke and K. Butts Pauly. MR thermometry. *J Magn Reson Imaging*, 27:376–390, Feb 2008. x, 35, 39, 49, 56, 73, 114, 124

[113] V. Rieke, K. K. Vigen, G. Sommer, B. L. Daniel, J. M. Pauly, and K. Butts. Referenceless PRF shift thermometry. *Magn Reson Med*, 51:1223–1231, Jun 2004. xi, 70

[114] V. Rieke, K. K. Vigen, G. Sommer, B. L. Daniel, J. M. Pauly, and K. Butts. Referenceless PRF shift thermometry. *Magn Reson Med*, 51:1223–1231, Jun 2004. 74, 112, 124

[115] M. Ries, B. D. de Senneville, S. Roujol, Y. Berber, B. Quesson, and C. Moonen. Real-time 3D target tracking in MRI guided focused ultrasound ablations in moving tissues. *Magn Reson Med*, 64:1704–1712, Dec 2010. 73

[116] J. Ritz. *Thermische In-situ Ablationsverfahren zur Behandlung von malignen hepatischen Tumoren - Experimentelle und klinische Untersuchungen zur Effektivitaetssteigerung und Therapieplanung*. Postdoctoral thesis Charite University of Berlin, Germany, 2006. 61

[117] Roggan. *Fortschritte in der Lasermedizin - Dosimetrie thermischer Laseranwendungen in der Medizin*. 1997. 31, 61, 95

[118] C. Rosenberg, R. Puls, K. Hegenscheid, J. Kuehn, T. Bollman, A. Westerholt, C. Weigel, and N. Hosten. Laser ablation of metastatic lesions of the lung: long-term outcome. *AJR Am J Roentgenol*, 192:785–792, Mar 2009. 15

[119] O. Rouviere, R. Souchon, R. Salomir, A. Gelet, J. Y. Chapelon, and D. Lyonnet. Transrectal high-intensity focused ultrasound ablation of prostate cancer: effective treatment requiring accurate imaging. *Eur J Radiol*, 63:317–327, Sep 2007. x, 15

[120] M. A. Rutherford, D. Azzopardi, A. Whitelaw, F. Cowan, S. Renowden, A. D. Edwards, and M. Thoresen. Mild hypothermia and the distribution of cerebral lesions in neonates with hypoxic-ischemic encephalopathy. *Pediatrics*, 116:1001–1006, Oct 2005. 35

[121] J. Sahuquillo and A. Vilalta. Cooling the injured brain: how does moderate hypothermia influence the pathophysiology of traumatic brain injury. *Curr. Pharm. Des.*, 13:2310–2322, 2007. 35

[122] R. Salomir, B. De Senneville, and C. Moonen. A Fast Calculation Method for Magnetic Field Inhomogeneity due to an Arbitrary Distribution of Bulk Susceptibility. *Concepts Magn Reson Part B (Magn Reson Engeneering)*, 19B:26–34, 2003. xii, 28, 74, 76

[123] R. Salomir, J. Palussiere, F. C. Vimeux, J. A. de Zwart, B. Quesson, M. Gauchet, P. Lelong, J. Pergrale, N. Grenier, and C. T. Moonen. Local hyperthermia with MR-guided focused ultrasound: spiral trajectory of the focal point optimized for temperature uniformity in the target region. *J Magn Reson Imaging*, 12:571–583, Oct 2000. 80, 93, 94, 111

[124] R. Salomir, M. Viallon, J. Roland, S. Terraz, D. Morel, C. Becker, and P. Gross. Reference-less PRFS MR thermometry using a thin open border and the harmonic functions theory: 2D experimental validation. *ISMRM Stockholm*, 247, 2010. 74, 109

[125] S. A. Sapareto and W. C. Dewey. Thermal dose determination in cancer therapy. *Int. J. Radiat. Oncol. Biol. Phys.*, 10:787–800, Jun 1984. 61, 69

[126] W. Schneider, H. Bernstein, and J. Pople. Proton magnetic resonance chemical shift of free (gaseous) and associated (liquid) hydride molecules. *Journal of Chemical Physics*, 28:601–607, 1958. 24

[127] N. F. Schwenzer, J. Machann, P. Martirosian, N. Stefan, C. Schraml, A. Fritsche, C. D. Claussen, and F. Schick. Quantification of pancreatic lipomatosis and liver steatosis by MRI: comparison of in/opposed-phase and spectral-spatial excitation techniques. *Invest Radiol*, 43:330–337, May 2008. 57

[128] R. J. Stafford, R. E. Price, C. J. Diederich, M. Kangasniemi, L. E. Olsson, and J. D. Hazle. Interleaved echo-planar imaging for fast multiplanar magnetic resonance temperature imaging of ultrasound thermal ablation therapy. *J Magn Reson Imaging*, 20:706–714, Oct 2004. 39

[129] R. Stollberger, P. W. Ascher, D. Huber, W. Renhart, H. Radner, and F. Ebner. Temperature monitoring of interstitial thermal tissue coagulation using MR phase images. *J Magn Reson Imaging*, 8:188–196, 1998. 77

[130] B. Taouli, A. Sandberg, A. Stemmer, T. Parikh, S. Wong, J. Xu, and V. S. Lee. Diffusion-weighted imaging of the liver: comparison of navigator triggered and breathhold acquisitions. *J Magn Reson Imaging*, 30:561–568, Sep 2009. 73

[131] B. A. Taylor, K. P. Hwang, A. M. Elliott, A. Shetty, J. D. Hazle, and R. J. Stafford. Dynamic chemical shift imaging for image-guided thermal therapy: analysis of feasibility and potential. *Med Phys*, 35:793–803, Feb 2008. 25

[132] S. Terraz, A. Cernicanu, M. Lepetit-Coiffe, M. Viallon, R. Salomir, G. Mentha, and C. D. Becker. Radiofrequency ablation of small liver malignancies under magnetic resonance guidance: progress in targeting and preliminary observations with temperature monitoring. *Eur Radiol*, 20:886–897, Apr 2010. 113

[133] S. A. Tisherman, A. Rodriguez, and P. Safar. Therapeutic hypothermia in traumatology. *Surg. Clin. North Am.*, 79:1269–1289, Dec 1999. 35

[134] M. Viallon, S. Terraz, J. Roland, E. Dumont, C. D. Becker, and R. Salomir. Observation and correction of transient cavitation-induced PRFS thermometry artifacts during radiofrequency ablation, using simultaneous ultrasound/MR imaging. *Med Phys*, 37:1491–1506, Apr 2010. 111, 113

[135] K. K. Vigen, B. L. Daniel, J. M. Pauly, and K. Butts. Triggered, navigated, multi-baseline method for proton resonance frequency temperature mapping with respiratory motion. *Magn Reson Med*, 50:1003–1010, Nov 2003. 70, 73, 112

[136] K. K. Vigen, B. L. Daniel, J. M. Pauly, and K. Butts. Triggered, navigated, multi-baseline method for proton resonance frequency temperature mapping with respiratory motion. *Magn Reson Med*, 50:1003–1010, Nov 2003. 124

[137] K. K. Vigen, J. Jarrard, V. Rieke, J. Frisoli, B. L. Daniel, and K. Butts Pauly. In vivo porcine liver radiofrequency ablation with simultaneous MR temperature imaging. *J Magn Reson Imaging*, 23:578–584, Apr 2006. x, 15, 35

[138] T. J. Vogl, M. G. Mack, P. Muller, C. Philipp, M. Juergens, D. Knobber, A. Roggan, P. Wust, V. Jahnke, and R. Felix. MR-guided laser-induced thermotherapy in tumors of the head and neck region: initial clinical results. *RöFo*, 163:505–514, Dec 1995. x, 35, 36

[139] T. J. Vogl, M. G. Mack, P. K. Muller, R. Straub, K. Engelmann, and K. Eichler. Interventional MR: interstitial therapy. *Eur Radiol*, 9:1479–1487, 1999. 35, 36

[140] M. von Siebenthal, G. Szekely, A. J. Lomax, and P. C. Cattin. Systematic errors in respiratory gating due to intrafraction deformations of the liver. *Med Phys*, 34:3620–3629, Sep 2007. 74

[141] C. Weidensteiner, N. Kerioui, B. Quesson, B. D. de Senneville, H. Trillaud, and C. T. Moonen. Stability of real-time MR temperature mapping in healthy and diseased human liver. *J Magn Reson Imaging*, 19:438–446, Apr 2004. 51, 52, 110

[142] C. Weidensteiner, B. Quesson, B. Caire-Gana, N. Kerioui, A. Rullier, H. Trillaud, and C. T. Moonen. Real-time MR temperature mapping of rabbit liver in vivo during thermal ablation. *Magn Reson Med*, 50:322–330, Aug 2003. 36

[143] Q. S. Xiang and F. Q. Ye. Correction for geometric distortion and N/2 ghosting in EPI by phase labeling for additional coordinate encoding (PLACE). *Magn Reson Med*, 57:731–741, Apr 2007. 50

[144] I. R. Young, J. W. Hand, A. Oatridge, and M. V. Prior. Modeling and observation of temperature changes in vivo using MRI. *Magn Reson Med*, 32:358–369, Sep 1994. 23, 26

[145] J. P. Yung, A. Shetty, A. Elliott, J. S. Weinberg, R. J. McNichols, A. Gowda, J. D. Hazle, and R. J. Stafford. Quantitative comparison of thermal dose models in normal canine brain. *Med Phys*, 37:5313–5321, Oct 2010. 34

[146] A. Zeiner, M. Holzer, F. Sterz, W. Behringer, W. Schorkhuber, M. Mullner, M. Frass, P. Siostrzonek, K. Ratheiser, A. Kaff, and A. N. Laggner. Mild resuscitative hypothermia to improve neurological outcome after cardiac arrest. A clinical feasibility trial.

Hypothermia After Cardiac Arrest (HACA) Study Group. *Stroke*, 31:86–94, Jan 2000. 35

[147] Q. Zhang, Y. C. Chung, J. S. Lewin, and J. L. Duerk. A method for simultaneous RF ablation and MRI. *J Magn Reson Imaging*, 8:110–114, 1998. 15, 35

Appendix

Abbreviations

α	Flip angle
α_E	Ernst angle
a	Temperature coefficient for the PRF method
$\vec{B}_0$	Static macroscopic magnetic flux density
$\vec{B}_1$	Macroscopic magnetic flux density induced by gradients
BHTE	BioHeat Transfer Equation
BW	Bandwidth
c_p	Specific heat capacity
CCC	Concordance Correlation Coefficient
CT	Computer Tomography
CNR	Contrast to Noise Ratio
δ	Shielding constant
D	Diffusion coefficient
ϵ	Average residual of the 2D phase's Laplacian needed for reference-less Temperature monitoring
$\vec{E}$	Electrical field
E_A	Activation energy
EPI	Echo Planar Imaging
ETL	Echo Train Length
ESP	Echo SPacing
φ	Phase of the MR image
φ_{Ref}	Reference phase

φ_T	Phase including the temperature dependent phase shift
FA	Flip Angle
FATSAT	FAT SATuration
FFT	Fast Fourier Transformation
FID	Free-Induction Decay
FLASH	Fast Low Angle SHot sequence
FOV	Field Of View
FT	Fourier Transformation / transform
γ	Gyromagnetic ratio
$\vec{G}$	Magnetic field gradients
GRAPPA	Generalized Autocalibrating Partially Parallel Acquisition K-space based parallel imaging technique
GRE	GRadient Echo
$\hbar$	Planck's constant
$\vec{H}$	Macroscopic magnetic field
H_z	Zeeman Hamiltonian
HIFU	High Intensity Focused Ultrasound
$\vec{I}$	Spin
$\vec{I}(x,y)$	Image intensity function
$\vec{j}_c$	Free current density
ICE	MR Image Calculation Environment
IR	Intersecting Region
k-space	All MR-data are acquired in the k-space ($\cong$frequency space). The Fourier transformation is used to transfer the measured MR-signal $S(k_x, k_y)$ within the k-space into an MR image $I(x,y)$ in the Cartesian coordinate system, which is monitored for clinical use.
$\vec{J}$	Angular momentum
κ	Boltzmann's constant
λ	Thermal conductivity coefficient

l_1, l_2	The l_1 and l_2 spaces are function spaces defined using natural generalizations of norms for finite-dimensional vector spaces (Lebesgue spaces).
LITT	Laser Induced Thermotherapy
LOA	Limits Of Agreement
$\vec{\mu}$	Magnetic moment
μ_0	Permeability of free space
m	Quantum number of I_z
$\vec{M}$	Magnetization
$\vec{M}_0$	Equilibrium Magnetization
MR	Magnetic Resonance
MRI	Magnetic Resonance Imaging
MRT	Magnetic Resonance Thermometry
NMR	Nuclear Magnetic Resonance
Ω	Damage Integral
OT	Overestimation of necrosis by TMap
PRF	Proton Resonance Frequency
PRFS	Proton Resonance Frequency Shift
Q	Energy
ρ	Tissue density
R, ρ	Radial coordinate and angular coordinate
RF	Radio Frequency
RFA	Radio Frequency Ablation
ROI	Region Of Interest
$\vec{S}(k_x, k_y)$	Frequency spectrum of the spin density distribution
SE	Spin Echo
SD	Standard Deviation
SDT	Standard Deviation of Temperature
seg EPI	Segmented Echo Planar Imaging

SENSE	SENSitivity Encoding
	Spatial space based parallel imaging technique
SNR	Signal-to-Noise Ratio
ss EPI	Single Shot Echo Planar Imaging
T	Temperature
TA	Time of Acquisition
T_1	Spin-lattice relaxation time
T_2	Spin-spin relaxation time
$T_2^\star$	Effective spin-spin relaxation time
TE	Echo Time
TMap	Temperature Map
TR	Repetition Time
ω	Frequency
ω_0	Resonance frequency called "Larmor frequency"
UT	Underestimation of necrosis by TMap
VIBE	Volume Interpolated Breath hold Examination
WIP	Work In Progress
χ	Susceptibility
ξ^2-test	Statistical hypothesis test

Curriculum Vitae

Personal Data

Name:	Antje Kickhefel
Date of Birth:	October 10, 1982
Sex:	Female
Nationality:	German
Degree:	Dipl. Phys.

Educational Qualifications

Since 2007	PhD student at the University of Tübingen and
	Siemens Healthcare H IM MR PLM AW Oncology
10/2002−09/2007	Study physics; field of study: Medical physics
	Ernst-Moritz-Arndt-University Greifswald
	Degree: Diploma in physics
	Mark: 1.2
10/2006−09/2007	Diploma thesis
	Ernst-Moritz-Arndt-University Greifswald
	Mark: 1.3
04/2005−09/2006	Research assistant
	Ernst-Moritz-Arndt-University Greifswald
	Institution for Diagnostic Radiology and Neurology
04/2005−09/2006	Research assistant
	Ernst-Moritz-Arndt-University Greifswald
	Hospital for Radiation therapy and Radio-oncology
10/2005−09/2006	Research assistant
	Ernst-Moritz-Arndt-University Greifswald
	Institution of physics

02/2005−03/2005	Advanced internship
	German Cancer Research Center Heidelberg
	Medical physics
	Radiation therapy
09/2003	Advanced internship
	Ernst-Moritz-Arndt-University Greifswald
	Institution for Diagnostic Radiology and Neurology
	Hospital for Radiation therapy and Radio-oncology
07/2003−07/2005	Member of student representatives of physics
09/1993−07/2002	Secondary school
	Lilienthal-Gymnasium Anklam
	Degree: Abitur
	Mark: 1.4

Awards / Stipends

2009	ISMRM Educational Stipend
2010	ISMRM Educational Stipend
2010	8th Interventional MRI Symposium: Cum Laude at Poster Award

List of Publications

Journal Publications

Antje Kickhefel, Christian Rosenberg, Joerg Roland, Patrick Gross, Fritz Schick, Norbert Hosten, Rares Salomir. Clinical evaluation of the near-harmonic 2D reference-less PRFS thermometry in liver using LITT ablation data. submitted to Europ Radiol

Antje Kickhefel, Clifford Weiss, Joerg Roland, Patrick Gross, Fritz Schick, Rares Salomir. Correction of Susceptibility-induced GRE Phase Shift for Accurate PRFS Thermometry proximal to Cryoablation Ice-ball. submitted to MAGMA

Antje Kickhefel, Joerg Roland, Clifford Weiss, Fritz Schick. Accuracy of Real-Time MR Temperature Mapping in the Brain: A comparison of fast sequences. Phys Med. 2010;26(4):192-201

Antje Kickhefel, Christian Rosenberg, Clifford R Weiss, Hansjoerg Rempp, Joerg Roland, Fritz Schick, Norbert Hosten. Real Time MR Temperature Monitoring of Laserinduced Thermotherapy in Human Liver. J Magn Reson Imaging. 2011;33(3):704-709

Rares Salomir, Magalie Viallon, Antje Kickhefel, Roland Joerg, Denis Morel, Thomas Goget, Lorena Petrusca, Vincent Auboiroux, Sylvain Terraz, Christoph D. Becker, Patrick Gross. Reference-free PRFS MR-thermometry using 2 near-harmonic 2D reconstruction of the background phase. submitted to Radiology

Jens-Peter Kühn, Matthias Evert, Antje Kickhefel, Katrin Hegenscheid, Gloger O, Lerch MM, Kühn J, Mensel B, Völzke H, Dombrowski F, Norbert Hosten, Ralf Puls. Noninvasive Quantification of Hepatic Fat Content Using Three-Point Dixon-MRI with a Correction for T2* Relaxation Effects. submitted to Radiology

Frank Hübner, Babak Bazrafshan, Jörg Roland, Antje Kickhefel, Thomas J. Vogl. The Influence of Nd:YAG Laser Irradiation on Fluoroptic Temperature Measurement: An Experimental Evaluation. submitted to J Magn Reson Imaging.

Hansjoerg Rempp, Petros Martirosian, Andreas Boss, Stephan Clasen, Antje Kickhefel, Markus Kraiger, Christina Schraml, Claus Detlef Claussen, Philippe Pereira, Fritz Schick. MR temperature monitoring applying the proton resonance frequency method in liver and kidney at 0.2 and 1.5 T: segment-specific attainable precision and breathing influence. Magn Reson Mater Phy 2008;21:333-343

Hansjoerg Rempp, Stephan Clasen, Andreas Boss, Joerg Roland, Antje Kickhefel, Christina Schraml, Claus Detlef Claussen, Fritz Schick, Philippe Pereira. Prediction of Cell Necrosis with Sequential Temperature Mapping after Radiofrequency Ablation. J Magn Reson Imaging. 2009;30(3):631-639.

Jens-Peter Kühn, Soenke Langner, Katrin Hegenscheid, Matthias Evert, Antje Kickhefel, Norbert Hosten, Ralf Puls. Magnetic resonance-guided upper abdominal biopsies in a high-field wide-bore3-T MRI system: feasibility, handling,and needle artefacts. Eur Radiol. 2010 May 26.

Patents

US Patentapplication 2010P13588US. Temperature measurement near an ice ball using a proton resonance frequency method and recalculation of susceptibility artifacts. Submitted

DE 2010E05711 DE/201015373. Verwendung von funktionalen Daten zur Registrierung von MR-Bildern (Selbstreferenzierung). Submitted

DE 2009E23969 DE/201007561. Single Species Multi-TE Signal Combine. Submitted

Conference Contributions

Talks

Antje Kickhefel, Joerg Roland, Fritz Schick. Real-Time Magnetic Resonance Temperature Mapping in the Brain. ISTU 9th International Symposium on Therapeutic Ultrasound Aix en Provence, France, 2009

Antje Kickhefel. Promovieren in der Industrie. Greifswalder Kollegforum: Medizin; Bildgebung in wissenschaftlichen Karrieren, Greifswald, Germany, 2010

Antje Kickhefel, Rares Salomir, Joerg Roland, Patrick Gross, Fritz Schick, Clifford Weiss. Temperature Measurement Nearby an Iceball using the Proton Resonance Frequency method: Recalculation of Susceptibility Artifacts. 18th Annual Meeting ISMRM Stockholm, Schweden, 2010

Antje Kickhefel, Christian Rosenberg, Joerg Roland, Patrick Gross, Fritz Schick, Norbert Hosten, Rares Salomir. Reference-less PRFS MR thermometry of the whole liver based on near-harmonic calculation: Clinical evaluation from LITT ablation data. 19th Annual Meeting ISMRM Montreal, Canada, 2011

Antje Kickhefel, Clifford Weiss, Joerg Roland, Patrick Gross, Fritz Schick, Rares Salomir. Correction of Susceptibility-induced GRE Phase Shift for Accurate PRFS Thermometry proximal to Cryoablation Ice-ball. Young Investigator Award, 28th Annual Meeting ESMRMB Antalya, Leipzig, 2011

Poster

Antje Kickhefel, Joerg Roland, Fritz Schick. Precision of Fast Temperature Mapping in Brain. 25th Annual Meeting ESMRMB Valencia, Spain, 2008

Hansjoerg Rempp, Andreas Boss, Antje Kickhefel, Petros Martirosian, F. Springer, Philippe Pereira, Fritz Schick. MR temperature monitoring applying the proton resonance frequency method in liver and kidney at 0.2 T and 1.5 T: Segment-specific attainable precision and breathing influence. 25th Annual Meeting ESMRMB Valencia, Spain, 2008

Antje Kickhefel, Joerg Roland, Fritz Schick. MR Temperature Mapping in Brain: A comparison of the GRE, segmented EPI and ss EPI sequences based on the Proton-Resonance-Frequency-Method. 17th Annual Meeting ISMRM Honolulu, Hawaii, 2009

Antje Kickhefel, Christian Rosenberg, Joerg Roland, Norbert Hosten. Patient study of real-time Magnetic Resonance Temperature Monitoring on moving organ simultaneous to Laserinduced Thermal Therapy (LITT). 17th Annual Meeting ISMRM Honolulu, Hawaii, 2009

Antje Kickhefel, Hansjoerg Rempp, Joerg Roland, Fritz Schick. Evaluation of Realtime Temperature Monitoring of human liver based on patients undergoing a Radiofrequency Ablation. 17th Annual Meeting ISMRM Honolulu, Hawaii, 2009

Antje Kickhefel, Eva Rothgang, Christian Rosenberg, Joerg Roland, Fritz Schick. Improving In-vivo MR Thermotherapy Reliability in Moving Organ by applying Pennes. 26th Annual Meeting ESMRMB Antalya, Turkey, 2009

Eva Rothgang, Antje Kickhefel, Joerg Roland, Christian Rosenberg , Joachim Hornegger , Christin H. Lorenz. Online improvement of the reliability of PRF based temperature maps displayed during laser-induced thermotherapy of liver lesions. 26th Annual Meeting ESMRMB Antalya, Turkey, 2009

Antje Kickhefel, Christian Rosenberg, Joerg Roland, Patrick Gross, Fritz Schick, Norbert Hosten, Rares Salomir. Evaluation of Reference-less PRF MR Thermometry using LITT patient data. 8th Interventional MRI Symposium, Leipzig, Germany, 2010

Rares Salomir, Magalie Viallon, Antje Kickhefel, Joerg Roland, Lorena Petrusca, Vincent Auboiroux, Thomas Goget, Sylvain Terraz, Denis Morel, Christoph D. Becker, Patrick Gross. Reference-free PRFS MR-thermometry using quasi-harmonic 2D reconstruction of the background phase. 8th Interventional MRI Symposium, Leipzig, Germany, 2010

Christian Rosenberg, Antje Kickhefel, Joerg Roland, Ralf Puls, Norbert Hosten. Continuous real-time MR-thermometry in moving organs Ü clinical routine during laser ablation of hepatic tumors. 8th Interventional MRI Symposium, Leipzig, Germany, 2010